How to Write A Request for Proposal (RFP) for a Telecommunications Cabling System

First Edition

D&D Incorporated

The contents of this publication reflect the views and opinions of the author acquired through many years of experience in the telecommunications industry. This publication is in no way providing legal or engineering professional services. The services of a professional person are recommended if legal or engineering services are needed. The publisher or author disclaims and will be held harmless from any loss or liability caused by utilization of the information provided herein.

Neither D&D Incorporated or the author are recommending any specific manufacturers cable or equipment, although, for illustration purposes it was necessary to specify manufacturers in the sample Request for Proposal.

ISBN 1-57820-002-6

About The Author

Don Bodnar, President of D&D Incorporated and author of this publication, has spent the last 15 years in the Telecommunications Cabling Industry. Since 1988 he has served as CEO of Mercury Communications Inc., a successful Telecommunications Cabling Contractor and Systems Integrator in the southern Connecticut region. He formed D&D Incorporated in 1993 to provide high level consulting services in the creation of Requests for Proposals (RFP's) for the telecommunications industry. In the early and mid 1980's he worked for several manufacturers of electronic systems in the capacity of procuring cabling contractors for installation of their various systems.

The history in the telecommunications industry described above has given Don the unique experience of having worked from all sides of the playing field. He has successfully played the role of customer, contractor and consultant for telecommunications cabling systems. This unique perspective is what compelled him to write this book. Having seen first hand the trials and tribulations of each discipline described above, Don feels he can offer valuable insight to others in this field.

To My Wife
and Partner
in Life,
Diane

Table Of Contents

Table Of Contents

Chapter One

Why Write an RFP

Objective

The intent of this book is not to teach cabling systems design or the legal requirements of any individual company with respect to contracting work. These parameters vary from application to application, from company to company and from local ordinance to local ordinance.

This book has been written with the intent of teaching the reader the mechanics of creating a structured and well defined Request for Proposal (RFP) document. Hopefully, by the end of this publication, the reader will have an understanding of the intricacies which are required in the creation of an RFP document for a communications cabling system. The ultimate goal of this exercise is to create a document that will solicit the best responses from prospective communications contractors.

Introduction

The bottom line in writing an RFP for a communications cabling system, is to obtain the best price at the highest level of quality possible for the services, labor and materials that are to be purchased. It should be noted that best price and highest quality are areas that, more often than not, conflict with one another, but we'll talk more about that later.

One of the first steps towards obtaining the above mentioned goals is creating an RFP which is clear concise and, most importantly, finely detailed from a cabling system design standpoint. A poorly defined cable system usually leads to one of the following conditions with regard to responses from bidders.

1. The bidder will inflate the bid price in order to cover itself for unknown expenses which inevitably arise during the installation process.

2. The bidder takes the bare essence of what the RFP document asked for in order to win the bid. Once the installation is under way, the owner is barraged with change orders for items not definitively specified in the RFP document.

3. The bidder provides low quality materials and inexperienced labor to perform the installation in order to submit a low price.

The second and third items above touch upon the price versus quality issue. What can be seen from the above dissertation is that a poorly defined cable system (sent out to bid) can ultimately and, most probably will, lead to higher costs. These costs are never usually evident at the start of the project, but whether its higher initial bids, change orders running amuck or reinstalling the system due to inferior workmanship and materials the costs are there.

Over the years this author has seen RFP's for multimillion dollar communications cable plants as well as projects as small as several hundred dollars. The RFP's associated with these projects have been as varied as the projects themselves. Some of these specifications would be as small as a few pages while others were hundreds of pages in length. This does not necessarily mean that the smaller RFP was associated with the smaller project or vice versa. Technical and contractual content of these documents have been as varied as their physical size. Some RFP's define hundreds of pages of contractual requirements while dedicating only minimal effort to actual system requirements. This type of a specification ultimately leads to one of the three contractor type of responses previously described.

Combining to many types of systems within one RFP is another mistake frequently made within this industry. Voice and data cabling have evolved to where economies are realized by installing these systems simultaneously. Video cabling can be added to the mix without much difficulty. Where the process begins to break down is when an end user starts to add ancillary equipment and systems to the RFP.

Paging and intercom systems are specialized industries and should be treated as such. This holds true for telephone switching equipment and computers. When all these requirements (and sometimes others) are tossed together into one RFP, the resulting document is often unclear, vague and contradictory. As explained above, we know what happens when a poorly defined specification is put out to bid. In this case the problem is further compounded by the number of systems the contractors are being asked to provide.

Another important point regarding multiple systems being put out in the same RFP concerns the abilities of the contractors themselves. A good communications contractor can provide and install voice and data cabling, active electronic hub equipment, video cabling and possibly some paging systems.

They are not usually, however, engaged in the direct sale of large telephone switching equipment, designers of broadband or baseband video head-end systems or in the retail computer business.

In such instances the contractor will usually partner with other companies or contractors in order to bid the entire project. The end user ultimately loses a large portion of control over their own project. Under this set of circumstances, the chances of disputes over systems requirements and a lot of finger pointing is high.

In conclusion, it is in this authors experience that keeping systems segregated and well defined will greatly increase the probability of obtaining the highest quality installation at the most cost effective price possible.

Chapter Two

Getting Started

The focus of this publication will be on the mechanics of creating a Request for Proposal for a Telecommunications Cabling System for voice and data communications systems. As indicated in Chapter 1, there is virtually no limit to the amount of contractual data which can be included as part of an RFP document, therefore it is always advisable to consult with a professional legal entity where questions may arise. Likewise, it is not the intention of this book to teach cabling system design. The information provided in the sample RFP is for the purpose of illustrating the steps required in creating a detailed Request for Proposal document. Invariably, the specific needs of any individual configuring a cabling system will be special to their particular environment.

The sample RFP provided at the end of this book will be used as a reference point for the outline to be discussed in chapter 3. This sample document defines the specifications for a voice and data communications cable plant for a hypothetical company known as the XYZ Corporation. This company is a campus of three buildings with copper and fiber optic communications links between each of the buildings. The cable plant includes cabling for voice and data communications systems.

A total of eight sections and six appendices comprise the sample RFP. After reading the document, it will become obvious that major emphasis has been placed on definition of the physical cabling system. The reasons for this have been mentioned earlier in chapter 1.

Before beginning a detailed discussion of the sections and sub-sections of the sample RFP, some general comments with regards to RFP writing should be noted.

The flow of information and how it is presented is almost as important as the information itself. RFP documents that bounce between technical and contractual requirements only tend to confuse bidding parties. This, as has been discussed, can lead to more costs in the long run. In addition, don't over specify certain requirements. An example of this would be copying a specification for the installation of a fiber optic ST connector into your RFP. This type of information becomes somewhat useless due to the fact that each manufacturer's installation requirements vary slightly and even change among different lines of the same product. This is not to say that discretion should be used when specifying the installation of a product that is new to the industry. In this case you may want to supply information on the manufacturer itself so bidders can contact them directly.

Another type of overkill found in RFP documents is with regards to cable specifications. Copying pages and pages of electrical and mechanical specifications, for a particular cable type, into an RFP only adds to clutter and confusion and is actually redundant information. This statement can be made for the following reasons:

1. The author of an RFP should specify a manufacturer and part number for the type of cable required.

2. Since the cable product is being specified by the RFP document, it is redundant to provide the actual technical specifications to bidding parties since they have no control over those specifications.

3. Specification and performance of testing procedures over the installed cable plant will verify the technical requirements.

To summarize the above point, since the RFP document should specify the manufacturers and the products being installed, there is no need to reiterate the technical specifications, of those products, in the body of the RFP. The best way to make sure that bidding parties are familiar with the products being specified is to perform the proper background checks into the responding vendors (this will be discussed in more detail throughout this book).

When writing an RFP it is vital to be clear, concise and to the point. Do not use conflicting or differing terminology while describing a particular item. An example of this would be calling a central wiring area a wiring closet in one section of the RFP and then referring to it as a communications room on drawings. This may seem a matter of semantics, but it can be very confusing to someone trying to figure out exactly what the requirements are. Always provide as much configuration information as possible including written descriptions of all aspects of the cabling system (i.e. horizontal cabling, vertical cabling etc.). Provide as many drawings as possible depicting not only physical cable plant topology but also rack and punch block layouts.

In addition to the above it is a good idea to shy away from tactics such as stating that the RFP is for information only and that it is the contractors responsibility to perform a design build of the project. In this authors opinion this is a "cover your rear" type of mentality that only serves to leave the entire project open to differing interpretations. This can only be a formula for problems during the course of the project. The point is, why write an RFP (or technical specification) if your telling your bidders to ignore it.

The more detailed the relevant information and the more defined the cabling system layout information is, the more precise the response to the RFP document will be. In addition, providing detailed and clear cabling system configuration information will ultimately reduce bidder inquiries and the inevitable addendums to the RFP.

While all attempts have been made to keep the language used in this book clear and concise, there is always the issue of semantics. That is, different terminology being used by different people to describe the same thing. A short list of terms and their definitions, as used in this book, is therefore presented below:

Central Computer Room - The area where computer servers and mainframes reside. This is also the location where backbone fiber optic cabling originates. The central computer room is often the same room as the phone switch room.

Communications Closet - An intermediate connection point for horizontal cabling, voice distribution cabling and data backbone cabling (see vertical cabling).

Cutover - The period of time when the cabling system is brought live and turned over to the customer.

Horizontal Cabling - Cabling which runs from the Communications Closet to the actual workstation locations.

MDF - Master Distribution Frame. A place where all voice distribution cable terminates and connects to the telephone switch. In many cases this term is generically used to describe a main distribution point for all voice and data backbone cabling.

Owner - A term used interchangeably in this book with "end user" to mean the recipient of the RFP responses and the cabling system.

Phone Switch Room - The room in which the telephone switch equipment and the MDF reside. This room is often the same room as the Central Computer Room.

Swing Area - An area in a building used to temporarily house employees being shifted around due to building renovations.

Vertical Cabling - Voice distribution and data backbone cabling. While this cabling might not necessarily be installed "vertically" in a building, it usually describes the cabling from the MDF and Central Computer Room to the Communications Closets.

Chapter Three

The Request for Proposal

This chapter will analyze the complete content of the sample RFP document provided at the end of this book. In addition, chapter 4 will cover some additional topics not included here.

The Cover Page

The cover page for your RFP should at a minimum include:

1. The company (end user) Name.

2. A statement identifying the document as a Request for Proposal.

3. A statement identifying what type of RFP the document is (i.e. communications cabling).

4. An issue date.

5. Identification of the preparer (company name or individual).

The following is additional information which can also be included on the cover page:

1. Bid or specification number.

2. Project number

3. Address of company (end user).

This is just enough information to let bidders know, at a glance, who their customer is and the type of project involved.

Table of Contents

A table of contents should be provided detailing all the sections, subsections and appendices of the RFP document. The major sections and subsections should be identified by page numbers and the appendices should be identified by each page of its content. As an example, if Appendix A includes all system drawings, then each of the drawing numbers and a description or drawing title should be listed.

The information provided in the table of contents (as described above) is important for the purpose of verifying that all bidders received all pages of the RFP. Should a drawing be listed in the table of contents, for example, it would be difficult for a bidder to argue (after the fact) that they never received the drawing.

Again, this is a way to alleviate possible back charges and finger pointing as the project progresses.

Section 1 - General Information

Subsection 1.1 - Introduction

A brief opening section describing the intent of the RFP document and also establishing the name, address, phone and fax number of the individual responsible for answering questions. This section should also define where the bid responses are to be delivered if different from the individual who is responding to questions.

This is a good place to specify whether or not questions regarding the RFP are to be submitted in writing.

Subsection 1.2 - Project Overview

General information regarding the cabling project should be included in this section. Items such as the physical address of the facility to be cabled and whether or not the facility is new construction, a renovation or an occupied environment can be mentioned here. In addition, this is a good spot to define the objectives of the project.

When defining project objectives stick to the goals of the newly proposed cable plant. Comments such as "the objective is to construct a new high speed backbone system" or "the goal of the project is to bring high speed communications capabilities to the desktop" are, in this authors opinion, acceptable. An example of what should not be included would be lengthy dissertation on the type of network software that may be used or discussions of past technology that is being scrapped. The RFP should be a specification for the proposed telecommunication cabling system only. Any information provided, which is not relevant to that purpose, tends to clutter and confuse the document.

Subsection 1.3 - Facility Layout

More detailed information regarding the facility to be cabled should be discussed in this section. The number of floors the project is to be spread over and the distribution of communications closets should be discussed here. Additionally, the location of the central computer and or phone switch room should be notated in this section.

A brief definition of the plan for the physical cabling can be identified in this section. Describe where horizontal cabling is coming from and going to for each floor. Likewise, definition of vertical cabling source and destination locations should be specified.

Lightning protection systems can be identified here, should there be any outside plant cabling involved with the project. With regards to outside plant cabling, include information on installation methods for the outside plant. As an example, describe whether or not the outside plant cabling will be performed aerially or underground. If the cabling will be underground, bidders will need to know if conduits are already in place or if new ones are to be installed. In addition, bidders will need to know what, if any, their responsibilities are regarding the installation of new conduits.

The same definition of requirements is required for any aerial cable work which might be required. Information should be provided on the overall plan for the aerial cable, who is providing telephone poles or if they are existing and if the aerial cabling traverses any publicly owned property.

Subsection 1.4 - Schedule

As much information as there is available about the project schedule should be presented here. Project start and completion dates should be discussed along with any major milestone requirements. As an example, if cable rough - in needs to be completed by a certain date (possibly due to the requirements of other trades on the project), then this should be specified.

Any installation work which may be required to take place simultaneously should also be addressed. For example, lets say the first and third floors of a facility are to be occupied and cutover first, with the second floor being occupied at a later date. Obviously it is critical that installation personnel begin work on these two floors first. It is vital that bidders are aware of facts such as these.

In addition to the above, if there are any plans to use an area of a facility as a temporary swing area, then scheduling of this swing area must be adequately defined.

Section 2 - Cabling System Design

Subsection 2.1 - Workstation Cabling

A detailed description of the exact workstation (horizontal) cabling requirements should be defined here. Cable information should include type and quantity of cables (i.e. 4 pair category 5, category 3, etc.) being installed to each workstation and jacket requirements (plenum rated, pvc, etc.). Termination methods for the workstation end and the communications closet end should be defined here. For example, description of the communications outlet at the workstation should include type of faceplate, type of jacks and mounting methods. Likewise, a description of the communications closet end should include type of patch panels or punch blocks to be used and their associated mounting methods. If mounting methods are to be different in different communications closets, then these issues also need to be defined.

Any variations of a typical communications outlet should also be discussed here. There may be some outlets which require an additional cable, for example. The additional cable and termination methods need to be identified and defined.

Additionally, this is a good area to describe any specific wiring standards which are required. This means to specify, for example, if you are utilizing 568A, 568B or any other wiring standard for your installation. Ultimately these standards will be dictated by the products which are specified later in the RFP, but it is still a good idea to be redundant on this issue.

Any customized termination requirements should also be identified here. As an example, if the intent is to split a 4 pair cable across two RJ11 connectors, then this information needs to be included.

At the end of each subsection under this section (Cabling System Design) any drawings which are included with the RFP document should be referenced.

Subsection 2.2 - Voice Distribution Cabling

Information supplied regarding the voice distribution (vertical) cabling should include the type of multipair cable being used (voice grade, category 3, etc.) and jacket requirements (plenum rated, pvc etc.). The number of cable pairs (100 pair, 200 pair, etc.) being installed from the MDF to each communications closet should be specified in detail. Termination equipment and mounting methods need to be accurately defined for each communications closet and the MDF area. This is a good place to comment on whose responsibility it is to install the plywood backboards for mounting punchblocks (if they are wall mounted). More often than not this issue gets forgotten until the project is well underway.

Should outside plant distribution cabling be included with the project (as is the case with the sample RFP), then explanation of how the voice distribution cabling is to interface with the outside plant cabling needs to be specified.

As a final note regarding voice distribution cabling, responsibility for performing workstation cross connects should be defined here. Normally the communications cabling vendor performs the cross connects in the communications closets while the telephone switch vendor performs the cross connects at the MDF. Although this is not etched in stone.

Subsection 2.3 - Fiber and Copper Backbone Cabling

In the sample RFP both fiber optic and copper cables are being installed for backbone (vertical) data cabling. Information for the fiber optic cable should include the number of strands being installed from the Central Computer Room to each of the communications closets, type of jacket required (plenum rated, pvc) and if any means of protecting the fiber is to be utilized. Specifically, is the fiber to be installed in innerduct. Should innerduct be required, then the type of material (plenum rated, fire retardent, etc.) needs to be defined.

Additionally, any inter-communications closet cabling should be fully defined in this section. This includes type of cabling, termination and termination hardware mounting methods. Inter-communications closet cabling may be cabling that is looped between communications closets for purposes specific to your applications.

Information defining the fiber optic cable as multimode, singlemode or a combination of both should also be supplied. The type of fiber optic connectors (ST, SC, etc.) to be used need to be stated along with termination hardware. Termination hardware includes fiber optic patch panels and the methods for mounting (rack mount, wall mount, etc.).

As with the fiber optic cable, full definition of any copper cabling used for backbone or data distribution purposes must be discussed. This includes cable type, termination methods (including any wiring standards), and termination equipment mounting methods.

Subsection 2.4 - Outside Plant Cabling

A brief reiteration from "Section 1.3 - Facility Layout" concerning outside plant cabling should be placed here. Specifically, requirements such as underground or aerial cabling and any pertinent data which may have an effect on the installation. As an example, if the outside plant cabling is to be in underground conduits, then bidders will need to know if those conduits are new, what size they are and if there is anything else currently installed within them. In addition, location of manholes or pull boxes for underground conduits should be identified. If an aerial solution is being used, then location and availability of telephone poles should be specified.

Source and destination points for all outside plant cabling needs to be established. If this was discussed in a previous section, it is a good idea to reiterate it here. Also, any special cable routing which may need to take place, once outside plant cabling enters a building, should be specified.

Once again, all components of the outside plant cabling need to be identified. For voice distribution cabling total pair counts need to be specified. Likewise total strand count and type (multimode or singlemode or both) for fiber optic cables. It should also be specified if additional protection for the fiber optic cable, by way of innerduct, is to be supplied. Information regarding the construction of the outside plant cable should be included. This means identifying if the cable is gel filled, armored, direct burial or any other type of construction.

Termination methods for both fiber optic and copper cabling should be specified. This information should include type of termination equipment for both the fiber optic and copper cabling along with mounting requirements.

Subsection 2.5 - Lightning Protection System

Under this section any lightning protection system being employed should be fully defined. This would include type of product, mounting methods and grounding requirements. Physical location of the lightning protection products should also be identified in this area.

Section 3 - Approved Manufacturers

This section provides information on all the manufacturers whose products are approved for the project specified in the RFP document. At this point, clarification needs to take place on the phrase “Approved Manufacturer”.

When specifying a cabling system, all the components need to be identified by manufacturer, part number and product description. The components defined in the RFP document (Section 4.0 - Product Specifications) should relate to the "Manufacturers of Choice" for the project at hand. Alternate manufacturers should be listed here along with the "Manufacturers of Choice" for purposes of competitive bidding. For example: Bidder 'A' may have a better price for category 5 cable utilizing a manufacturer that is not specified as the primary "Manufacturer of Choice", but is still an approved manufacturer of category 5 cable.

In the sample RFP the manufacturers name is supplied along with address and phone number. This is probably overkill and simply supplying the name of the manufacturer should be sufficient.

To maintain order and continuity it is a good idea to separate manufacturers by their various product lines. This tact has been followed in the sample RFP with manufacturers being segregated into the categories listed below:

Subsection 3.1 - Copper and Fiber Optic Cable

Subsection 3.2 - Termination Hardware

Subsection 3.3 - Lightning Protection

Subsection 3.4 - Miscellaneous Hardware

Miscellaneous hardware can include items such as cable hanging hardware, firestop material and equipment racks.

Section 4 - Product Specifications

This section of the RFP document specifies the "Manufacturer's of Choice" for the products to be included in the cabling system. Every component to be used in the cabling system should be defined here. For purposes of clarity, the manufacturer and product part number should be specified along with a description of where the product is to be used in the cabling system. As an example, if you were specifying the manufacturer and part number of a category 5 horizontal station cable you should also include a description of the product and information on how that product is going to be terminated at each end.

Once again, products should be segregated into their various groups as specified in section 3. The following items are defined in the sample RFP.

Subsection 4.1 - Copper and Fiber Optic Cable

Copper Outside Plant Cable

Fiber Optic Outside Plant Cable

Copper Voice Distribution Cable

Copper Data Distribution Cable

Multimode Fiber Optic Distribution Cabling

Singlemode Fiber Optic Distribution Cabling

Copper Workstation Cable

Subsection 4.2 - Termination Hardware

Workstation Faceplates

Workstation Connectors

Workstation Surface Boxes

Copper Patch Panels

Various Fiber Optic Cabinets

Multimode Fiber Optic Coupler Panels

Singlemode Fiber Optic Coupler Panels

Multimode Fiber Optic Connectors

Singlemode Fiber Optic Connectors

Punch Blocks

Connecting Blocks

Subsection 4.3 - Lightning Protection

Lightning Protector Panels

Lightning Protection Modules

Subsection 4.4 - Miscellaneous Hardware

Equipment Racks

Wire Management Panels

Innerduct

Punch Block Mounting Bars

Firestop Material

Cable Tray

Wall Duct

Subsection 4.5 - Notes to Product Specifications

This is a good spot to place a reminder that only the "Products of Choice" or approved equivalents are acceptable materials for the project. Although it will be mentioned in the next section, a comment stating that product specification sheets will be required can be placed here.

Section 5 - Requirements

Instructions to bidders on all aspects of the installation process along with contractual requirements need to be included in this section and the following subsections.

Subsection 5.1 - Installation

Statements relating to what is expected of the winning bidder with reference to the cabling system should be identified here. Comments such as "all materials are to be installed in accordance with manufacturers specifications" can be placed here.

Specific mounting requirements for various components should be discussed in this subsection. Any areas that might be considered vague need to be clarified here. The sample RFP, for example, states in an earlier section that ladder rack will be used in the communications closets. This is vague in that a definite configuration is not specified. This comment in this section clarifies the issue by stating that "For the purposes of this bid assume 20 feet of ladder rack is to be installed per communications closet".

Subsection 5.2 - Labeling

Direction on where labels are to be placed and the specific type of labels to be used need to be defined in this section. It should also be specified that only machine generated labels are acceptable (should that be the case). A statement that communications cabling system labeling must conform to EIA/TIA 606 labeling standards can be inserted here (more on the EIA/TIA later in this chapter and in chapter 5).

Subsection 5.3 - Testing

Testing is an area that is open to extreme overkill and one which can cost serious money if over specified. This author has seen RFP documents which require, for example, a test report, from the manufacturer, for every reel of cable that is delivered to the job site. This is not generally what cable manufacturers supply with their products (at least not at the time of this writing). Cable is manufactured on large master reels, which are tested, and then spooled to smaller reels. Requiring a test report for each reel of cable costs money and someone has to pay for that. The same holds true for fiber optic cable.

Another point to be made regarding testing is in the area of printed test results for horizontal workstation cabling. RFP writers will often require the installers to provide a printed test report for every cable tested. Depending on the size of the project, this can be thousands of test reports. The reports, once printed, usually get put on a shelf somewhere and are never looked at.

Certification scanners on the market today, create nice neat test reports for every cable which is tested. These reports are usually a page in length, are stored electronically or magnetically and can be read by most word processor or spreadsheet programs. Once the test reports are in a spreadsheet or word processor format, specific test reports can be searched by the software engines included with these programs.

For precisely these reasons it is in this authors opinion that printing test reports are a waste of time and money. Issues such as these are precisely why we have computers.

The bottom line in cable system testing is to insure the installed communications cabling system operates to the specifications defined by the RFP document. These specifications are defined under this subsection and under subsection 5.6, Codes and Standards.

For the purposes of illustrating nominal testing procedures, the cabling system provided in the sample RFP document can be divided into the following sections:

1. Category 3 Workstation Cable - Since this cable, by specification of the RFP, will be split into a non-standard configuration (splitting the 4 pair cable between two jacks) and used for voice or analog circuits only, a simple continuity and polarity test is all that is required.

2. Category 5 Workstation Cable - In this example, the category 5 workstation cable will adhere to EIA/TIA 568B Standards (as defined by the products specified) therefore, sophisticated certification testing should be called for. The sample RFP specifies the exact test results to be supplied and the type of test equipment (or equivalent) to be utilized.

3. Category 3 Voice Distribution cable - As with item number 1 above, simple continuity and polarity testing is all that is required.

4. Category 5 Data Tie Cabling - Testing for this cable will be the same as for item number 2 above.

5. Multimode Fiber Optic Distribution Cable - Power meter testing should be sufficient for most multimode applications. Testing bandwidth(s) should be specified along with whether the testing should be single or bi-directional. Also, in this instance, maximum loss values should be specified. Various manufacturers test gear provide for electronic or magnetic storage of data thereby relegating the need for printed media.

6. Singlemode Fiber Optic Distribution Cable - As specified in item number 5 above, power meter testing should be sufficient for most multimode applications. Testing bandwidth(s) should be specified along with whether the testing should be single or bi-directional. Also, in this instance, maximum loss values should be specified.

Several points need to be made regarding category 5 certification testing, fiber optic testing and the now often mentioned point of overkill. This author has seen the testing sections of many RFP's include rewrites of the actual values required for certification. For example, "Cable length not to exceed 100 meters". This kind of information can become long and tedious and in this authors opinion only tends to clutter up the document.

Category 5 cabling has become such an industry wide standard that exact specifications are available everywhere.

Also, since specifications for category 5 cabling systems are frequently made more stringent, the most current specification is what will be desired for the cabling system under discussion. The only time exact values should be included with an RFP is if the requirements do not conform to industry wide standards. As an example, if environmental conditions dictate some sort of custom specification, then this should be stated in the document.

With regards to fiber optic testing, many RFP's specify the use of an OTDR (Optical Time Domain Reflectometer) for testing purposes. This is a more costly form of measurement and the most significant information that will be derived from this piece of equipment (outside of loss measurements) will be cable length. Most fiber optic distribution cabling, within a building, probably averages around 400 feet. More accurate measurements can be derived from architectural drawings. Fiber Optic cable can usually run to a distance of about 2 kilometers without needing any active electronics. The point is, what is the cost versus information benefit of testing with an OTDR and is it worth it.

This is not to say that the OTDR is not an excellent diagnostic tool for finding breaks or other aberrations within a fiber optic strand. Although, if a quality fiber optic product is used and proper installation methods are employed, these problems seldom, if ever, arise in short fiber optic cable installations

Subsection 5.4 - Documentation

Documentation for communications cabling systems usually consists of "As - Built" drawings and the presentation of the testing data defined in subsection 5.3.

As - built drawings contain information regarding cable routing, workstation outlet numbering and any other information which may pertain to the physical cabling system. This documentation can be provided as information hand drawn on architectural prints or it can be required in some sort of CAD (Computer Aided Design) format. If CAD drawings are required, then background drawings in a magnetic media (floppy disk) will need to be supplied to the installation contractor so the proper information can be added. The magnetic form of these drawings should be provided in a file format popular with many CAD programs (such as dxf or dwg).

Test results for the cabling system should be required in magnetic format wherever possible. A paper log sheet identifying every workstation number and the cabling installed to it should also be required. This can be a simple sheet with a column used for a reference number, a column used for workstation identification number and a column for each cable installed to the workstation. There are two purposes for this log sheet, first it keeps a running total of the exact workstation count and secondly it identifies every workstation test report that is supplied in magnetic format. The test log sheet also acts as a quick check of testing performed but not supplied in magnetic format. For example, in the sample RFP only continuity and polarity testing are required for the voice workstation cabling. In addition, the 4 pair voice cable is being split between two jacks. Due to this configuration there will be no test report generated, therefore the test log sheet can be used to verify that testing was performed on those cables.

Any test reports supplied via magnetic media should be supplied in a popular format that is readable by most software. Providing data in ASCII files is one possible format.

Subsection 5.5 - Submittals

Product specification sheets for every item used in the cabling system should be supplied by the installation contractor prior to project execution. This will verify compliance with products specified by the RFP document or any approved alternates which are being utilized. The product specification sheets should contain information defining the manufacturer, product and part number for every component of the cabling system.

Subsection 5.6 - Codes and Standards

Many agencies exist which govern the installation requirements for telecommunications cabling systems. Compliance with these organizations is required if the cabling system is to meet electrical, transmission, Radio Frequency (RF) emission and fire code standards. Some of the pertinent agencies are listed below. Chapter 5 supplies more in - depth information on these agencies and how to get in touch with them.

ANSI - American National Standards Institute

EIA/TIA - Electronic Industries Association Telecommunications Industries Association

FCC - Federal Communication Commission

IEEE - Institute of Electrical and Electronic Engineers

NEC - National Electric Code (The NEC is a publication of the National Fire Protection Association)

NEMA - National Electrical Manufacturers Association

NFPA - National Fire Protection Association

OSHA - Occupational Safety and Health Administration

UL - Underwriters Laboratories

Subsection 5.7 - Warranty

Many manufacturers in conjunction with installation contractors are offering 15 to 20 year warranties on the installed telecommunications cabling system. Some of these warranty agreements are for the physical cable and connector products while other agreements cover the "applications" which are to run [transmit] over the cabling system. Since manufacturers of cable and connectors are much larger organizations (usually) than installation contractors, the theory is greater customer satisfaction due to the backing of a larger company. While this may be somewhat true there is another point to consider here. Most of these warranty packages are written with the stipulation that only the manufacturers supplying the warranty and / or their strategic business partners products can be used in the cabling system. Should another manufacturers product be introduced into the cabling system, lets say five years down the road, then in all probability the original warranty will become void. This means if you use any other manufacturers product in your system (at any time) you have probably voided your warranty.

The problem with all of this, is that the owner looses a great degree of flexibility with regards to which products can be used with the installed telecommunications cabling system.

Another important point to be made, with regards to manufacturer warranty issues, relates to installation contractor qualifications. Most installation contractors are unable to become certified installers of every manufacturers products. Unfortunately, what happens is that good installation contractors can only align themselves with a few manufacturers. One reason for this is because most manufacturers requires a specified dollar amount that each contractor must purchase annually in order to maintain a "Qualified Installer" status. Obviously, every manufacturer wants every installer to focus on their product.

As can be seen in the above dissertation, is if you adamantly specify a specific manufacturers warranty program, in your RFP document, you are severly limiting your current and future options.

Most manufacturers require installation contractors to go through rigid training programs before they can become "Qualified Installers" of their products. With this in mind, part of the due diligence, in finding reputable installation contractors, might be to find out which manufacturers they are aligned with and for how long the partnership has been in existence. This information can then be used as a gauge in determining the quality of the installation contractor.

The whole point of this discussion is that if reputable installation contractors are used and quality products from reputable manufacturers are specified, then the need for one of these manufacturer warranty programs may not exist. Many companies only require a (1) year warranty on workmanship and material. Other companies require (5) year warranty packages.

The lesson to be learned here is that having the confidence in the chosen manufacturers products and the installation contractors workmanship enables the RFP writer to tailor a warranty package to the specific requirements of the project. What is gained is freedom from the encumbrances and inflexibility's of manufacturer warranty programs.

Subsection 5.8 - Changes and Alternates

This article sets the rules for any deviations (if any) from the specified telecommunications cabling system which would be acceptable in the bidders responses. These deviations could include everything from alternate products to telecommunications cabling system design changes.

In addition, this area can also be used to set the guidelines for the structure of the response itself. For example, if it is required that pricing be presented in exactly the format defined in the RFP document, then it should be stated here.

It must also be stated what the consequences of not adhering to guidelines presented here would be. Such as, any deviations from the pricing format will be considered non-responsive.

Subsection 5.9 - Project Management

It is a good idea to specify how the project is to be managed once it is awarded and underway. Consideration must be given to the qualifications of the person to be appointed project manager from the installation contractor. Information about the project manager should include experience with projects of the size being specified and with the type of products being installed.

What will be required of the project manager should also be specified in this section. As an example, if weekly progress reports or attendance at regular construction meetings are required, then it should be stated here. The decision making authority of the project manager should also be defined. Does the person have the authority to commit additional manpower for example.

A final point regarding project managers relates to longevity. It should be clearly stated that the project manager appointed by the installation contractor cannot be changed. There may be certain situations where this is unavoidable, but changing a project manager mid-project can usually lead to problems.

Subsection 5.10 - Right of Refusal

This qualification defines the intentions or rights of the owner with regards to the RFP responses. Some points to be notated here can include the following:

1. The right to reject any or all responses.

2. The right to enter into negotiations with more than one qualified bidder.

3. The right to award the project to other than the lowest bidder.

Basically, any rights which are in the best interest of the owner should be stated here.

This is also a good section to insert a statement that specifies RFP response preparation costs as being the responsibility of the responding parties (bidders).

Subsection 5.11 - Work by Others

When installing a telecommunications cabling system there are areas of work scope that may overlap with other working trades. In new construction, for example, installing an electrical back box for the workstation outlet may be covered by this RFP or the task may be part of the electrical specification. Installation of plywood backboards (as mentioned earlier) may be required or it may be covered under the carpenters contract.

Any areas of possible overlap should be defined and responsibility for the task in question must be identified. Aside from the above examples, the majority of these areas of overlap will involve cabling pathways and will include, but not be limited to, items such as cable tray, conduit stub-ups, power poles and floor poke thrus.

Subsection 5.12 - Insurance's

Requirements for insurance coverage vary from company to company. At a minimum, it should be required that a copy of the respondents insurance certificates be presented with the RFP response. At a maximum, actual insurance ceilings required for specific categories of coverage should be defined. The following is a list of typical insurance coverage's, but always check with all the parties concerned (building owner, company management, etc.) for any specific insurances which may be required.

Type of Coverage	Typical Limit
Commercial General Liability	
General Aggregate	$2,000,000.00
Products Comprehensive	$2,000,000.00
Personal Injury	$1,000,000.00
Each Occurrence	$1,000,000.00
Fire Damage (any one fire)	$500,000.00
Medical Expenses (any one person)	$20,000.00
Automobile Liability	
Combined Single Limit (any auto)	$500,000.00

Excess Liability (Umbrella Coverage)

Each Occurrence	$5,000,000.00
Aggregate	$5,000,000.00

Workers Compensation & Employers Liability *

Each Accident	$100,000.00
Disease - Policy Limit	$500,000.00
Disease - Each Employee	$100,000.00

* Statutory limits for workers compensation insurance vary from state to state. Check with the state in which the project is to take place for exact coverage requirements.

Subsection 5.13 - Debris Removal

A simple article stating that the contractor is responsible for their own debris clean-up on a daily basis. Back charges can occur from several sources if a contractor does not clean up after itself. The owner of the telecommunications system may be charged from the general contractor or the building owner (if a leased property) for debris removal if responsibility for this issue is not properly defined.

Subsection 5.14 - Material Handling

This article declares that all cost associated with "handling of materials" is the responsibility of the contracting party. The word "handling" can have several meanings. This authors definition of "handling" includes, but is not limited to, getting materials to the job site, storage of materials at the job site and transportation of materials from the storage area to the work locations.

It should also be noted, under this section, that material delivery schedules required to meet installation completion dates are the responsibility of the contracting party.

Subsection 5.15 - Subcontractors

In many cases a contractor (who may already be a subcontractor) will require the assistance of a subcontractor. For example, a general contractor may subcontract a project to a telecommunications contractor who in turn may subcontract a portion of their contract to another contractor (or subcontractor). To further this example, the telecommunications contract may call for installation of underground cabling between several buildings. This may require the trenching and laying of conduit between the facilities in question. Many telecommunications installation companies may retain the equipment to pull cable in this environment, but they may not posses the trenching gear. This will then require the services of a subcontractor to perform the trenching.

What must be specified here, is who takes responsibility for the actions of the subcontractor. Following the above example, lets assume the trenching subcontractor does not perform the proper research regarding the area in which the trenching is to take place. Lets further assume, in this hypothetical example, that the trenching subcontractor begins work and cuts into a telephone company fiber optic line that feeds a substantial number of telephone company customers.

The obvious costs associated with this mishap would be lost revenue to the telephone company due to phone system downtime and the costs associated with the physical repair of the cut fiber optic cable. Some not so obvious liabilities might include claims filed from telephone company customers due to the fact that they could not contact emergency services in time of crisis. Whatever the case may be, it should be evident from this example that the possibility of litigation due to the poor workmanship of unscrupulous subcontractors can be a costly affair.

The whole purpose of this subsection is to specify the responsibility, for the actions of, any subcontractors involved in the project. Essentially it should be stated that the actions of any subcontractors are the responsibility of the prime contractor. Assistance from a legal professional may be required in the exact wording of an article, such as this, which would be acceptable to the end user.

Subsection 5.16 - Compliance

It is a good practice to require RFP respondents to include a compliance summary with their responses. A compliance summary is detailed acceptance of every section and subsection of the RFP document. The basic format for a compliance summary requires the respondents to create an outline which relates to every article of the RFP document and declare that they (the respondent) either complies with or takes exception to the article. Should an exception be taken, then a full explanation must accompany the exception.

Requiring a compliance summary serves the following purposes:

1. It insures that each respondent has read and understands the full scope of the RFP document and the project.

2. Any exceptions which a contractor may identify are clearly defined and can be used as an analytical tool in rating competing contractors.

A compliance summary can only serve to hinder possible disputes as a project gets underway. It clearly identifies that the contracting party has read and understands the full scope of the work and each article detailed in the RFP document.

Section 6 - Pricing

Subsection 6.1 - Pricing Format

Defining how pricing is to be presented from bidding contractors is critical to assuring "apples to apples" responses. This subsection should include or reference a sample pricing format that all bidding parties are to follow. A statement identifying the consequences of not following the required format should also be placed here.

The sample RFP provided in appendix 'A' of this book outlines just such a pricing format. The following items are specified in the sample pricing sheets:

1. Reference Number
2. Description (of installation task)
3. Manufacturer
4. Part Number
5. Quantity
6. Unit Material Cost
7. Unit Labor Cost
8. Extended Cost

9. Section Total

Additional columns that can be added are "Extended Labor" and "Extended Material" costs for each line item. It may also be a requirement to know the total "Material" and the total "Labor" costs for the project. One possible reason for needing this information is in the area of local tax laws. The state of Connecticut, for example, (as of this writing) requires sales tax to be charged for labor and materials for any project that is not a new construction. Sales tax is charged on materials only for projects that are new construction. The accounting department of the company purchasing the telecommunications cabling system may also need this pricing information for bookkeeping purposes.

While on the subject of taxes, responsibility for sales or any other type of applicable tax should be defined here. For example, does the purchasing party require taxes to be included with each line item, is a lump sum line item for taxes acceptable or do taxes even need to be broken out at the time of the bid responses. These tax requirements need to be clearly specified in order to insure bid responses that are equivalent in all aspects.

After analysis of the pricing format provided with the sample RFP (appendix 'B' of the sample RFP), it should be evident that this author takes a slightly different approach to the information provided in the RFP document. The industry standard approach to the bidding process usually takes the following course:

1. An architect generates architectural drawings.

2. Either the end user generates an RFP document or the end user hires a consultant for this task.

3. The architectural drawings are sent to bidders with the RFP document.

4. It is then the bidders responsibility to take the information off of the architectural drawings. This information includes horizontal workstation counts, cable footages, etc..

5. The bidder must then formulate all the quantities of the piece parts required for the cabling system.

The approach this author believes in, is providing bidders with the information contained in items 4 & 5 listed above. This will require a little more work on the part of author of the RFP document but, it will insure bid responses that are absolutely equal in content and pricing. Obviously, a confidence level must exist, with the author of the RFP, that the quantities of materials specified are correct.

Specifications for quantities of items such as faceplates, jacks and patch panels are usually a fixed number based upon workstations counts. Unfortunately, the most costly item is the one that can fluctuate the most with regards to quantity. This item is the labor and material required to install the raw cable. Many variables come in to play when determining cable footages, the single most important being the physical pathway from the cables source to its destination.

For the reason stated above, if the approach taken in the RFP document is to specify quantities, then the following caveats should be placed in this section:

1. It should be stated that it is the responding parties responsibility to verify and confirm all specified cable footages.

2. It should also be stated that it is the responding parties responsibility to identify any omission, made by the author of the RFP document, which may constitute additional costs to the end user with regards to the telecommunications cabling system being proposed.

Subsection 6.2 - Unit Prices

Telecommunications cabling projects are dynamic environments which continually change. It has been this authors experience that, on average, projects normally increase in size from 25% to 30% between start and completion. Given the fact that the cabling system will inevitably increase in size, it is only prudent to require unit prices from the responding bidders for certain aspects of the project.

The unit prices required here are in addition to those supplied in the base project pricing. These can be prices for items such as patch cords, fiber optic jumper cables, and the total cost of a standard workstation drop.

Subsection 6.3 - Labor Rates

During the course of a large project the need usually arises for work to be performed on a "time and material" basis. This situation can surface when work tasks are to be performed which are outside of the scope of the original project.

Labor rates from the responding parties should include "Straight Time", "Over Time" and "Double Time" prices. In addition, definition of the precise times with which these rates apply should be included with the response.

Subsection 6.4 - Invoicing Format

The sample RFP includes a suggested invoicing format (appendix 'E' of the sample RFP) for the winning installation bidder to provide with milestone billing. This format is similar to the pricing format with the addition of several columns to the spread sheet. These columns are as follows:

1. Previous Complete
2. Current Complete
3. Previous Billed
4. Current Billed

When this format for invoicing is utilized correctly, it offers precise control over each item of the installation. A quick glance at this type of invoicing spreadsheet defines exactly how much of the project has been implemented, how much and which tasks are left to implement and how much the project has cost to date.

Section 7 - Bid Schedule

This section should specify any dates which are pertinent to the RFP document and responses. The date, time and place of the bid conference can also be defined here. This is also a good place to specify how many copies of the bid response are required to be submitted.

A point has to be made here regarding the time frame between the issuance of the RFP document and the response due date. Adequate time must be allowed for bidders to generate an appropriate response. Obviously, the time allowed for responses should vary proportionately with the size of the project. Requiring unreasonable time frames for bid responses only strengthens the possibility of receiving sloppy and inadequate proposals.

Section 8 - Bidder Qualifications

Subsection 8.1 - References

It is a good idea to solicit references of past projects from responding bidders. This is part of the procedure in the verification of the credentials of the responding parties, although this author would not put a tremendous amount of importance on this process. This statement can be made for the following single most important reason:

> **"Has anyone ever seen a solicitor of any type give a reference that would be derogatory to the solicitors cause".**

When checking on bidders, a better idea is to use competing bidders references against one another. For example, bidder 'A' and bidder 'B' each supply three references, While conversing with the references of bidder 'A' try and find out if they have ever heard of, or have any information regarding bidder 'B' and vice versa. While the telecommunications industry is huge, local installation communities tend to be small and well networked. Basically, everyone knows everyone. Therefore information regarding competing bidders should be abundant, just ask around.

The following items are some sample topics which should be touched upon when interviewing contractor references:

1. Was the project completed on schedule?

2. How did the contractor handle any problems which may have surfaced?

3. Was there adequate manpower to complete critical path milestones?

4. Was the project manager knowledgeable and attentive to the needs of the end-user?

5. What was the disposition of the installation staff? Were they courteous and professional?

6. How did the installation contractor interact with other trades on the project?

7. Where there any questionable change orders submitted?

8. Would the interviewee select the installation contractor for future projects, if the opportunity arose?

9. Has the interviewee ever heard of or done business with any of the competing bidders being checked out?

10. If the answer to question #9 is yes, then does the interviewee know of any customers which might be familiar with the work of the competing bidders?

There will probably be other questions needed to be added to this list, which are particular to the project being proposed but, this is a good starting point. Obviously, leads to any additional information obtained from these interviews should be thoroughly investigated.

Subsection 8.2 - Resumes

Requiring resumes of several of the installation contractors project managers is prudent due diligence work. These resumes should include any licenses or accreditations held by the project managers, project experience, years with the installation contractor and years in the industry. This information will help build a picture of the overall experience and capabilities of the bidding contractor. For example, if the responding contractor has only one experienced project manager, the question arises as to how the proposed project will be handled should the contractor in question have another project running simultaneously.

This is a good area to reiterate a point discussed in subsection 5.9 - Project Management. The point was, that once a project manager is chosen, that individual cannot be removed from the project.

There is one other issue to be made under this subsection. It is not a good idea to ask responding contractors for resumes on the exact personnel who will be staffing the proposed project. It is in this authors experience that many installation contractors will not choose an internal project manager until the project is formally awarded. From the standpoint of the installation contractor, there are many variables which must be weighed when appointing project managers and installation personnel. Current workload, projected workload, project schedules and manpower availability are only a few of the items a contractor must consider.

In addition to the above, the position "Project Manager" means different things to different companies. In very large contracting companies a project manager is usually an individual who interfaces with the project from the standpoint of material delivery scheduling, providing documentation and supporting the project foreman. In small to medium size companies a project manager may be a working foreman who interfaces with his office for the services of material procurement and documentation generation.

Considering the above, and the fact that many projects will not start until months after the RFP has been responded to, illustrates the uselessness of requiring installation contractors to provide exact project staffing information with the RFP response.

It is this authors opinion that much more valuable insight into an installation contractor can be gained by viewing a roster of resumes for several or all of their project managers.

Subsection 8.3 - Organizational Structure

A brief outline of the organizational structure of the responding contractors is another piece of information which will help to put together an accurate picture of their organizations. Organizational charts which provide positions in the responding company, without attached names, are pretty much worthless. The most valuable information would be a listing of the positions and names of key personnel such as principals, administrative, designers, project managers and the current number of installation technicians on staff.

Once again, this information will give insight into the size and capabilities of the responding contractors. It will also provide a loose escalation path for any problems which may not be properly handled by the appointed project staff.

Subsection 8.4 - Licensing

In many states licenses are required for all installation personnel, who are physically performing the cabling system installation. In addition, the contracting company may be required to hold a valid license as a contractor in the state in which the work is being performed.

As an example, the State of Connecticut requires different classes of licenses based upon the voltage levels being transmitted across the installed cabling. An 'E2' license holder can work on cabling circuits up to high voltages (normal electrical type of wiring) while a 'T2' license holder can only install cabling carrying voltages of up to 48 Volts. A 'T2' license, in this example, is adequate for most voice and data cabling systems. The license types listed above, though, are only valid for the physical installation work. In order to be a "Licensed Contractor" (in Connecticut) the company would need a person with an 'E1' or 'T1' license (for low voltage cabling) on staff.

In light of the previous example, it is prudent to check with the state in which the installation is to take place and find out what licenses may be required. This can be done by talking to the building department, usually located in the town hall, of the town or city who has jurisdiction over the installation. It can then be required, of responding contractors, to provide verification of their ability to perform work in the locality in which the project is to take place.

Another point which must be discussed, relevant to the issue of licensing for telecommunications cabling installers, is in the area of electrical permits. Many states require "Electrical Permits" to be taken out with the local governing authority prior to commencing work. The "Permit" is a vehicle by which the local town or city can check on the installation work and make sure it is compliant with national and local electrical installation codes. For more information on permits, see chapter 4 of this book entitled "Some Final Thoughts".

A very important point must be made regarding licensing and the example illustrated here. The licenses specified, in this example, are State of Connecticut requirements which insure proper installation of building cabling for fire and safety considerations. These licenses do not encompass the realm of certification for high speed data transmission. The point is, do not confuse government licensing requirements with certification programs for high speed data cabling offered by other agencies or institutions.

Subsection 8.5 - Manpower

Almost inevitably, the requirement will exist for a high concentration of manpower to complete some critical path portion of any sizable cabling installation project. It is therefore good to know where the responding contractor is going to draw this manpower from when the time arises.

It must be noted that some leverage is due a responsible contractor in providing manpower to perform critical path items of a project. Some installation contractors can be very creative when manipulating manpower to accomplish specified project milestones. As an example, if the contractor does not have the available manpower during a normal shift to complete an item, they might be willing to work a crew a second shift, at no charge to the owner, in order to accomplish the task at hand.

A thorough investigation of the responding contractors and, their history in the industry, can only serve to enhance or detract from the confidence level given to the submitted responses of question such as those asked by this subsection.

Appendix 'A' - Drawings

As mentioned in chapter 2, information flow is critical to producing an understandable RFP. Nowhere is this more important than with presentation of drawings. Mixing a rack layout drawing in between a riser diagram and building layout will only tend to confuse the picture that is trying to be presented.

A top down view is one that should be employed in the presentation of drawings. This means to start with the biggest picture and progressively break the system down into smaller and smaller pieces. The sample RFP supplied with this publication provides a set of (17) drawings. These drawings start with a broad overview of the campus, then move on to the building level followed by a complete distribution system overview and ending with rack and punchblock level drawings.

This author would strongly suggest the use of color for riser diagrams or for any drawings which depict cable routing. Inexpensive color ink jet printers are readily available and can produce excellent quality drawings. The color drawings included in the sample RFP were generated in just this fashion.

While there is no standard (as of this writing) stating which colors are to be used for different aspects of the cabling system, this author normally uses the color red to depict fiber optic cabling, blue for voice cabling and green for data cabling.

The following discussion will look at each drawing, of the sample RFP, in more detail and identify some of the information which should be included in each. It should be noted that each drawing should, at a minimum, contain the following information:

1. Project Name

2. Drawing Name

3. Drawing Number

4. Date Drawn

5. Revision Number (and date of revision)

Drawings change, more often than not, before project completion. It is therefore very important to keep accurate track of any revisions made to installation drawings and have a system in place to assure the revisions have been distributed to the installation contractor. Performing an installation from an incorrect revision of a drawing can sometimes be a very costly proposition.

Drawing # Sample - 01 / Campus Layout

A very broad overview of the sample campus environment showing the relative locations of the three buildings. Some general information is provided showing the pathways (underground conduits) which will be utilized to connect the buildings.

Many times, architectural site plans are available for contractors to take off exact measurements. Unfortunately, these plans are not always present, therefore a drawing such as this serves as a guide for taking field measurements.

Drawings # Sample - 02, 03, 04 / Building Riser Diagrams

These drawings are a combination of building riser diagrams and cabling system one - line drawings. A building riser diagram depicts a physical representation of the building being cabled with some cabling pathway information included. This can be illustrated by viewing drawing "Sample - 03" which shows workstation cabling from the basement level being terminated in the first floor communications closet.

A cabling system one - line drawing represents cabling information which is "Typical" to the entire cabling system. An example of this is illustrated in the drawings being discussed in this section. A single workstation is depicted as being associated with each communications closet but, the cabling shown is typical of every workstation being installed to that closet.

The three drawings being discussed here represent the next step in the top down view of the cabling system being analyzed. Information presented by these drawings is still slightly broad but, much more detailed then the previous drawing. Cable routing, termination requirements, distribution system information and building information are all items which are illustrated by these drawings. As much information as possible should be included on these drawings. Any pertinent information with reference to the physical building should also be included on these drawings. This can be information describing the relative locations of the communications closets or anything that might be considered a non - typical installation environment.

Drawings # Sample - 05 Distribution System Overview

The distribution system described in the sample RFP is somewhat complex and requires further definition to create a clear understanding of its scope. Once again, in the top down presentation model, this drawing takes the level of detail one step further than the previous drawings.

This drawing illustrates a detailed overview of the topology of the cabling distribution system. It defines precise cable pair and fiber optic strand counts between the computer room, the building entrances and each communications closet. As described in the beginning of this section, color creates a vivid representation and segregation between voice, copper data and fiber optic cabling.

Drawings # Sample - 06 Through Sample 17

These drawings illustrate the final step in the top down model used in this section. They provide detailed information with respect to equipment rack layout, punch block arrangement, and fiber optic cabinet placement. This information is provided for every communications closet, building entrance area and the computer room.

The level of detail provided in these drawings literally include descriptions of every component in the cabling system. Rack layouts, as an example, need to show exact placement and quantities of the hardware being installed into them. The drawings should be an exact representation of what the equipment racks are to look like at the completion of the project. This level of detail also holds true for any wall mounted fields including punch block fields, lightning protection fields or wall mountable fiber optic patch fields.

The drawings described in this section will, for the most part, be the blueprints by which the installation contractor quotes and builds the cabling system. Obviously, the need for clear, concise and non - conflicting information is vital to the implementation of a successful project. In addition, the more detailed the information, the less questions there will be surrounding the RFP document and the accompanying responses.

Appendix 'B' - Pricing Format

The pricing format for this hypothetical cable system is broken into five sections. Detailed unit costs are required for each of the three buildings and the outside plant cabling along with a spreadsheet summarizing the costs of the first four sections.

See subsection 6.1 "Pricing Format" presented earlier in this chapter for further information on pricing.

Appendix 'C' - Unit Pricing

See subsection 6.2 "Unit Pricing" presented earlier in this chapter for information on unit pricing.

Appendix 'D' - Labor Rates

See subsection 6.3 "Labor Rates" presented earlier in this chapter for information on labor rates.

Appendix 'E' - Invoicing Format

See subsection 6.4 "Invoicing Format" presented earlier in this chapter for information on invoicing formats.

Appendix 'F' - Workstation Schedule

The workstation schedule defines the total number of workstation cabling drops to be quoted under the RFP document. This schedule follows the philosophy of providing the bidding contractors all the pertinent information, including material quantities, required to submit accurate and equal responses. As mentioned in subsection 6.1 of this chapter, this requires a little more work on the part of the RFP writer but, it insures absolute control over all aspects of system pricing and ultimately over the entire project.

Chapter Four

Some Final Thoughts

As discussed in chapter 1, RFP documents are as varied in size and content as the individuals who write them. If someone in the legal profession were charged with the task of writing an RFP for a telecommunications cabling system, it would probably be filled with page after page of contractural requirements with only minimal importance being placed on cabling system design. This author has seen many RFP documents of this nature which, most of the time, are generated by government agencies and institutions. What happens is, a boilerplate RFP document is continually added to year after year after year. Ultimately, less and less attention is paid to the actual cabling system then the contractual requirements. Unfortunately, the end user suffers for the reasons discussed in chapter 1. If the end user happens to be a government agency or institution then everyone suffers.

Conversely, if the RFP were to be written by a strict engineering type of professional, then the document would probably be overkilled with unnecessary technical specification of the type discussed throughout this book. In this instance the RFP document would almost assuredly lack even the most minimal of contractual requirements.

The intent of this publication is to present a middle ground when writing an RFP document. Hopefully, it is evident from this book the importance this author places on detailed design information being the predominant focus of the RFP document. This is not to diminish the need for clear contractual requirements as this author believes has also been presented here. As stated previously, these contractual needs vary from organization to organization therefore it is best to consult a legal professional regarding the specific requirements for any particular project.

Ultimately, the more relevant detail provided with the RFP, the more accurate the responses will be and the more control the end user will have over the project. This is, after all, what it's all about. Additionally, a clearly written RFP document will eliminate or reduce to a minimum any uncertainty on the part of the responding installation contractors.

For one reason or another, several issues, ideas and opinions have not been covered in chapter 3 or the sample RFP document. The following pages of this chapter will discuss these remaining items in detail although in no specific order.

Permits

The issue of permits was briefly touched upon in subsection 8.4 of chapter 3 entitled licensing, this issue will now be discussed more thoroughly. Telecommunications cabling usually falls under the jurisdiction of a municipalities electrical department. The type of permit required therefore, is an electrical permit. If the state in which the installation is to take place requires an electrical permit for the project, then any costs associated with permit must be included in the respondents pricing. This cost can be simply added as a line item to the pricing format spreadsheet.

Costs for permits vary widely from municipality to municipality but they can range anywhere from $10.00 to $30.00 or more per $1,000.00 of the telecommunications contract value. Some municipalities will not charge for the permit if a building permit has been issued.

<u>Bonding</u>

There are several types of bonds which are often required from an installation contractor for telecommunications cabling projects. Two types of these bonds are as follows:

Bid Bond - This type of bond insures the end user that the price submitted by the RFP respondents is valid and will remain that way for some specified period of time. These bonds are usually issued for approximately 5% to 10% of the price submitted by the bidder. These bonds are required to be submitted with the bid responses.

Performance Bond - The performance bond insures the end user that the system to be installed will be done so in accordance with the plans and specifications of the RFP document. These bonds are issued at the total value of the contract price and are usually required before the start of the project.

While bid bonds do not usually cost the bidding contractors anything, the performance bond can be expensive. These bonds can range in price anywhere from 1.5% to 3% of the total contract value with a sliding scale taking effect after some value ceiling. As an example, it may cost a contractor 2.5% for the first $100,000.00 of contract value with the rate lowering to 1.5% for any contract value over the $100,000.00 amount.

Bonds are a form of insurance, to the end user, that the system being procured will be installed correctly and to specification. Bonds, like insurances, are usually a very subjective matter. Essentially, does the cost justify the security being provided by the bond. It should also be noted that the bond really does not guarantee a perfect installation, what it provides is a legal means for the end user to take action against an unscrupulous contractor via the bonding company. There have been many bonded jobs, improperly installed, which results in litigation between the bonding company, the end user and the contractor. This type of scenario usually turns into a legal nightmare lasting years.

The decision to require bonding is one that must be made by the end user and the RFP writer. At a minimum, the responding contractors should be required to provide information on their bonding companies and the limits to which they are bondable. It is in this authors opinion that one of the best ways to protect against poor workmanship is a thorough evaluation of responding installation contractors. Checking on the bonding companies is another resource for performing this due diligence.

Labor Unions

Contracting installation work with a contractor who is affiliated with a labor union is a very important consideration for many companies. Two of the major unions which perform telecommunications installation work are:

1. The International Brotherhood of Electrical Workers (IBEW)

2. The Communications Workers of America (CWA)

Both of these labor organizations are affiliated with the AFL/CIO (American Federation of Labor / Congress of Industrial Organizations) and have extensive experience with telecommunications cabling systems.

This author has more experience with the practices and policies of the IBEW therefore the discussion will focus on this labor union.

Several years ago the mere mention of a union contractor performing a telecommunications cabling system installation would make an Information Systems manager cringe. This mentality was prevalent due to a lack of confidence in union workmanship and an outright fear of unions. Over the past several years unions have had their own shakeout along with corporate America. What this has produced, at least in the case of the IBEW, is an organization that is very cognizant of the huge size and potential of the telecommunications cabling industry. This realization has caused some changes to take place (from the perspective of an installation contractor) within the IBEW. These changes include:

1. Regular training of members in the installation of telecommunications cabling products.

2. Labor rates which are more competitive with non-union shops.

3. More of a partnership than an adversarial relationship with signatory contractors.

The basic structure of the IBEW creates "Local" pools of labor all over the United States. It can also be reasonably assured that this labor is properly licensed should the cabling installation be in a state requiring licensing. Given these facts, the statement can be made that an end user can feel confident that plenty of qualified (or licensed) manpower will be available if the choice is made to contract with an IBEW contractor.

The Bid Conference

When organizing a bid conference and choosing the number of bidders which will respond to the RFP, it is in this authors experience that it is in the best interest of the end user to keep the number of bidders in the range of 5 to 7. This is not possible for government bids which must be open to the public but, the government bidding process will be used to illustrate the point.

Many bid conferences have been attended by this author, for public bids, which literally take place in school auditoriums filled with bidding contractors. The problem with this is a lot of the reputable contractors, attending these conferences, many times throw up their hands in disgust and no - bid the project not wanting to waste resources on a project they feel they do not have a chance of winning.

To take this a step back, one of the reasons reputable contractors back out of a bid, as in this example, is due to the inadequate nature of the RFP documents supplied. Inadequate RFP's create the problems described in chapter 1 of this book. Unfortunately, the inadequate RFP's are many times created due to the same bidding process that is being discussed here.

The point is, keeping the number of competing bidders to a reasonable level will only serve to enhance the competitive bidding process. Contractors will feel they have a reasonable chance of winning the project which will almost assuredly be reflected in their pricing.

Construction Schedules

The conditions set forth by many RFP's require an installation contractor to provide project schedules and Gantt charts describing the exact time frames which will be required to complete a project. This is a rather ridiculous requirement, especially if the project is a major renovation or a new construction. This is due to the fact that the project schedule of the telecommunications cabling installation is dependent upon almost every other construction trade on the project. Cable cannot be roughed in until studding is installed, communications closet equipment can not be installed until the room is built, final termination and testing of workstation drops can not take place until modular furniture is installed (which almost always happens last).

A more relevant question would be to ask what are the total man hours estimated for the project and what are the estimated manpower loading projections. These questions will return answers that can be analyzed and used in determining the creditability of the responding contractor.

Grounding Systems

Changing the direction of this chapter, momentarily, some important points must be made regarding grounding systems. Grounding systems were not mentioned in chapter 3 or the sample RFP due to the fact that many times the responsibility for the installation of the grounding system is under the realm of the electrical contractor.

While the EIA/TIA specifies requirements for the grounding of telecommunications termination equipment, it does not specify whose responsibility it is to install the grounding system. The current division of labor (between the electrical contractor and the telecommunications contractor) is settling upon the electrical contractor bringing a ground connection point into a communications closet, computer room or anyplace a ground connection is required. The telecommunications contractor then makes the connection between this ground and the telecommunications termination equipment to be grounded. This, of course, is not written in stone.

A Changing Industry

A few final points have to be made with respect to the changing dynamics of the telecommunications industry. These changes include not only a steady flow of new products coming to market but, more importantly, new and evolving standards which continue to reshape the telecommunications cabling business.

Chapter 5 of this book presents a listing of agencies and organizations whose regulations and standards have a direct influence on the telecommunications cabling industry. Also listed are some of the articles of these regulatory agencies which, at the time of this writing, define the parameters of various aspects of telecommunications cabling installation practices.

The point to be made here is that standards are continually being redefined which result in changes to the articles, of these agencies, which govern installation practices. Being aware of the latest documentation releases, from these agencies, is the only way to insure compliance with the most current standards.

This same philosophy holds true for new telecommunications cabling products which become available on almost a daily basis. A continual education of new products on the market is essential to configuring telecommunications cabling systems which are standards compliant and which fit the most efficient and cost effective installation standards of the day.

An example of the continual introduction of new products into the marketplace is illustrated in the following: The sample RFP defines a particular vertical wire manager for use in the wire management of the hypothetical cabling system. Since the writing of this book, new products have come to market (from the same manufacturer) which are more suited to the function of vertical wire management. This is not to say that the specified product is obsolete or does not perform the function for which it was specified, only that a more efficient method of vertical wire management has been established.

Unfortunately, the continual changes in the availability of products such as these could not be incorporated into this book. This is due to the fact that the book would be in a constant state of rewrite and would never get published.

Chapter Five

Governing Agencies

Chapter five presents a listing of agencies governing different aspects of the telecommunications cabling industry. The listings and the articles presented are by no means conclusive or an end all in sources of information available. In addition, as previously mentioned, standards are continually being redefined therefore some of the articles listed here may be superseded by the time this book is read.

One of the best ways to interact with these organizations is on the World Wide Web of the Internet. An enormous wealth of information is available over this media.

ANSI - American National Standards Institute

11 West 42nd Street
New York, NY 10036

Telephone - (212) 642 - 4900
Fax - (212) 398 - 0023

Web Site - http://www.ansi.org/home.html

Relevant Standard - X3T9.5 FDDI Standard

BICSI - Building Industry Consulting Service International

10500 University Center Drive
Tampa, Florida 33612 - 6415

Telephone - (813) 979 - 1991

Web Site - http://www.bicsi.org/

Relevant Document - Telecommunications Distribution Methods Manual

EIA/TIA - Electronic Industries Association / Telecommunications Industries Association

The articles presented here are the result of a collaborative effort between these two organizations.

EIA
2500 Wilson Boulevard
Arlington, VA 22201

Web Site - http://www.eia.org/

TIA

No main address is available.

Telephone - (703) 907 - 7700

Web Site - http://www.industry.net/tia

The TIA organization is segmented into the following divisions:

• Fiber Optic Division - Concentrates on business issues which impact the use of fiber optics technology.

• Mobile Communications Division - Supports the needs and interests of manufacturers of portable and vehicular two-way radio equipment.

• Network Equipment Division - This divisions three sections address policy and technical issues related to intelligent networks, point to point microwave and broadcast transmission.

• Satellite Communications Division - The SCD will oversee a structure of sections and committees which address issues specific to various sectors of the satellite industry.

• User Premises Equipment Division (UPED) - The activities of this division center around regulatory changes at the Federal Communications Commission (FCC) which are related to the safety and performance standards of customer premises equipment.

• Roadrunners International - This is an industry service organization, sponsored and funded by TIA, which focuses on the human interest issues important to the telecommunications industry.

Relevant Standards:

EIA/TIA 568 Commercial Building Wiring Standard

EIA/TIA 568 TSB-36 Twisted Pair Cable Categories

EIA/TIA 568 TSB-40 Twisted Pair Connecting Hardware Categories

EIA/TIA 568 TSB-67 Transmission Specifications for Unshielded Twisted Pair Cables

EIA/TIA 568 TSB-72 Centralized Optical Fiber Cabling Guide

EIA/TIA 569 Commercial Building Standard for Telecommunications Pathways and Spaces Standards (Proposal SP2950).

EIA/TIA 606 Administration Standard for Telecommunications Infrastructure of Commercial Buildings.

EIA/TIA 607 Commercial Building Grounding and Bonding Requirements for Telecommunications Standard.

EIA/TIA 526 Series Documents on Optical Fiber Systems Testing.

EIA/TIA 455 Series Documents on Fiber Optic Test Procedures.

FCC - Federal Communications Commission

1919 M Street N.W.
Washington D.C. 20554

Telephone (202) 418 - 0200

Web Site - http://www.fcc.gov

Items covered by the FCC are developed by one of seven operating bureaus and offices categorized by substantive area.

- The Common Carrier Bureau handles domestic wiring telephony.

- The Mass Media Bureau regulates television and radio broadcasts.

- The Wireless Bureau oversees wireless services such as private radio, cellular telephone, personal communications services (PCS) and pagers.

- The Cable Services Bureau regulates cable television and related services.

- The International Bureau regulates international and satellite communications.

- The Compliance & Information Bureau investigates violations and answers questions.

- The Office of Engineering & Technology evaluates technologies and equipment.

Relevant Documents - FCC Title 47

FCC Part 15

FCC Part 68

IEEE - Institute of Electrical and Electronic Engineers

Operations Center
445 Hoes Lane
Piscataway, NJ 08855-1331

Telephone (908) 981 - 0060

Ask IEEE Document Delivery

Telephone (800) 949 - IEEE
Fax - (415) 259 - 6045
e-Mail: askieee@ieee.org

Conference Information

Telephone (908) 562 - 3878
Fax - (908) 981 - 1769
e-Mail: conference.services@ieee.org

Customer Service

Telephone (800) 678 - IEEE
Fax - (908) 981 - 9667
e-Mail: customer.service@ieee.org

Standards Information

Telephone (908) 562 - 3800
Fax - (908) 562 - 1571
e-Mail: stds.info@ieee.org

Web Site - http://www.ieee.org/i3e_hp.html

Relevant Documents -

IEEE 802.5

IEEE 802.3

<u>NEMA</u> - National Electrical Manufacturers Association

1300 North 17th Street
Suite 1847
Rosslyn, VA 22209

Telephone - (703) 841 - 3200
Fax - (703) 841 - 3300

Web Site - http://www.nema.org/

<u>NFPA</u> - National Fire Protection Association

1 Batterymarch Park
P.O. Box 9101
Quincy, MA 02269 - 9101

Telephone (800) 344 - 3555

Web Site - http://www.wpi.edu/~fpc/nfpa.html

Relevant Documents -

NFPA 70 - National Electrical Code

NFPA 75 Protection of Electronic Computer Equipment

OSHA - Occupational Safety and Health Administration

No main address is available.

Office of Information and Consumer Affairs
Telephone (202) 219 - 8151

Office of Statistics
Telephone (202) 219 - 6463

Office of Field Programs
(202) 219 - 7725

Web Site - http://www.osha.gov/

Relevant Documents - OSHA 29 CFR 1926/1910

UL - Underwriters Laboratories

333 Pfingsten Road
Northbrook, IL 60062

Telephone - (847) 272 - 8800
Fax - (847) 509 - 6249

Global Engineering Documents

2805 McGaw Avenue
Irvine, California 92714

Telephone - (800) 854 - 7179

Copies of EIA/TIA wiring standards can be obtained from this organization.

Sample RFP

Request for Proposal

XYZ Corporation

Communications Cabling Infrastructure

Issue Date xx/xx/xx

CONTENTS

1.0 General Information

1.1 Introduction

The XYZ Corporation is, by way of this RFP, soliciting vendors for pricing to furnish and install a communications cable plant as defined in the following pages. All responses and questions regarding this RFP should be submitted in writing to the following:

Your Name
Your Title
XYZ Corporation
Main St.
Anytown, USA 00000-0000
Telephone - (000) 000 - 0000
Fax - (000) 000 - 0000

1.2 Project Overview

The XYZ Corporation currently occupies two facilities, within the same campus, in Anytown, USA. These facilities are located at 1 Main St. and 3 Main St.. A third facility at 2 Main St., which is currently unoccupied, will be taken over by the XYZ Corporation. Underground conduit will be constructed between these three facilities (see dwg# Sample-01) which will include conduits for communications cabling.

The objective of this project is to install a new cable plant in each facility along with outside plant cabling to provide a communications link between each building.

1.3 Facility Layout

Building 1. - A three level building with one communications closet on each level. The second and third floor closets are stacked near the center of the building while the first floor closet is offset to one side of the building and is not directly under either of the other closets. The first floor closet will also serve as the building entrance room. All workstation cabling will be served from the closet on its respective floor. The building entrance room (first floor closet) will serve as the termination point for the inter-building distribution cabling and the lightning protection system.

Building 2. - A three level building consisting of a basement level, first floor and second floor. First and second floor communications closets are stacked near the center of the building. Workstation cabling on the basement level will be fed from the first floor communications closet. All other workstation cabling will be served from the closet on its respective floor. Building entrance areas, for this building, have not been defined as of this date. For bidding purposes it should be assumed that these areas will be located on the basement level within 30' of where each new underground conduit link accesses the building. The building entrance areas will serve as the termination point for the inter building voice distribution cabling and the lightning protection system. The computer room for the entire campus will be located at one end of the first floor of this building. The telephone switch and the voice MDF for the campus will also be located in this room.

Building 3. - A two level building with a communications closet on each level. The first and second floor closets are stacked near the center of the building. All workstation cabling will be served from the closet on its respective floor. The building entrance area, for this building, has not been defined as of this date. For bidding purposes it should be assumed that this area will be located on the first floor within 30' of where the new underground conduit link accesses the building. The building entrance area will serve as the termination point for the inter building voice distribution cabling, the lightning protection system and the distribution point for the inter building fiber optic cabling.

Underground Conduit - Each conduit link will have available (4) 4" conduits for communications cabling purposes.

1.4 Schedule

The first building to be cabled will be Building 2 with work to start the beginning of June 19XX. The building underground conduit links will be constructed in the summer of 19XX with a projected completion of August. Cutover of Building 2 will take place in September of 19XX at which time all outside plant cabling must be completed. A new phone switch and computer equipment, which will reside in the computer room of Building 2, will be brought on line at this time to service all three facilities.

Once the above installation has been successfully completed, work will begin at the Building 3 facility with an anticipated completion of November of 19XX. The last facility to be installed will be Building 1 with an expected completion of February 19XX.

Installation work at Building 1 and Building 3 will be accomplished in a phased manner. A swing area will be defined for XYZ Corporation personnel as each area is being installed.

2.0 Cabling System Design

2.1 Workstation Cabling

Building 1 - Each workstation in this facility will be cabled with (2) 4 pair category 5 plenum rated cables and (1) 4 pair category 3 plenum rated cable. All workstation cabling will terminate at the workstation in a single gang outlet consisting of (1) dual RJ45 category 5 module, (1) dual RJ45 category 3 module and (1) blank module. Each 4 pair category 5 cable will terminate on (1) RJ45 category 5 jack. The 4 pair category 3 cable will be split with 2 pairs going to each category 3 jack. All faceplates will be single gang. Modular furniture faceplates will be mounted to a single gang surface box and screwed to the modular furniture at floor level.

In the communications closet, category 5 workstation cabling will terminate on 48 port RJ45 category 5 patch panels which will be mounted in 19" free standing racks. The 4 pair category

3 cable will terminate on 300 pair 110 style punch blocks. Punch blocks in the first floor communications closet will be wall mounted. In the second and third floor closets the 110 punch blocks will be rack mounted.

Reference drawing #'s Sample-01, Sample-02, Sample-06, Sample-07, Sample-08, Sample-11, Sample-13

Building 2 - Each workstation in this facility will be cabled with (2) 4 pair category 5 plenum rated cables and (1) 4 pair category 3 plenum rated cable. All workstation cabling will terminate at the workstation in a single gang outlet consisting of (1) dual RJ45 category 5 module, (1) dual RJ45 category 4 module and (1) blank module. Each 4 pair category 5 cable will terminate on (1) RJ45 category 5 jack. The 4 pair category 3 cable will be split with 2 pairs going to each category 4 jack. All faceplates will be single gang. Modular furniture faceplates will be mounted to a single gang surface box and screwed to the modular furniture at floor level.

In the communications closet, category 5 workstation cabling will terminate on 48 port RJ45 category 5 patch panels which will be mounted in 19" free standing racks. The 4 pair category 3 cable will terminate on 300 pair 110 style punch blocks. Punch blocks in the first and second floor communications closets will be wall mounted.

Reference drawing #'s Sample-01, Sample-03, Sample-09, Sample-11

Building 3 - Each workstation in this facility will be cabled with (2) 4 pair category 5 plenum rated cables and (1) 4 pair category 3 plenum rated cable. All workstation cabling will terminate at the workstation in a single gang outlet consisting of (1) dual RJ45 category 5 module, (1) dual RJ45 category 4 module and (1) blank module. Each 4 pair category 5 cable will terminate on (1) RJ45 category 5 jack. The 4 pair category 3 cable will be split with 2 pairs going to each category 4 jack. All faceplates will be single gang. Modular furniture faceplates will be mounted to a single gang surface box and screwed to the modular furniture at floor level.

In the communications closet, category 5 workstation cabling will terminate on 48 port RJ45 category 5 patch panels which will be mounted in 19" free standing racks. The 4 pair category 3 cable will terminate on 300 pair 110 style punch blocks. Punch blocks in the first and second floor communications closets will be wall mounted.
Reference drawing #'s Sample-01, Sample-08, Sample-09, Sample-11, Sample-12

2.2 Voice Distribution Cabling

Building 1 - Voice distribution cabling for this facility will originate from the lightning protection system located in the building entrance room. From this position (400) cable pairs will be distributed to the first floor communications closet (located in the same room as the building entrance), (700) cable pairs will be distributed to the second floor communications closet and (500) cable pairs will be distributed to the third floor communications closet.

All voice distribution cable pairs will terminate on 300 pair 110 style punch blocks in each communications closet. Punch blocks in the plaza level communications closet will be wall mounted. In the first and second floor closets the 110 punch blocks will be rack mounted. Voice distribution cabling in the building entrance room will terminate on the lightning protection system.

Reference drawing #'s Sample-02, Sample-05, Sample-11, Sample-13

Building 2 - Voice distribution cabling for this facility will originate from the computer room MDF (Main Distribution Frame) located on the first floor. From this position (500) cable pairs will be distributed to the first floor communications closet and (500) cable pairs will be distributed to the second floor communications closet. In addition, (1600) cable pairs will run from the computer room MDF to the building entrance area lightning protection system feeding Building 1 and (1200) cable pairs will run from the computer room MDF to the building entrance area lightning protection system feeding Building 3.

All voice distribution cable pairs will terminate on 300 pair 110 style punch blocks in each communications closet and at the computer room MDF. Punch blocks in the first and second floor communications closets will be wall mounted. All punch blocks in the computer room MDF will be rack mounted. Voice distribution cabling in the building entrance areas will terminate on the lightning protection system.
Reference drawing #'s Sample-03, Sample-05, Sample-11, Sample-14

Building 3 - Voice distribution cabling for this facility will originate from the lightning protection system located in the building entrance area. From this position (600) cable pairs will be distributed to the first floor communications closet and (600) cable pairs will be distributed to the second floor communications closet.

All voice distribution cable pairs will terminate on 300 pair 110 style punch blocks in each communications closet. Punch blocks in the first and second floor closets will be wall mounted. Voice distribution cabling in the building entrance area will terminate on the lightning protection system.

Reference drawing #'s Sample-04, Sample-05, Sample-11, Sample-12

2.3 Fiber and Copper Backbone Cabling

Building 1 - Fiber optic backbone cabling for this facility will originate from the fiber distribution patch panel located in the building entrance room. From this position (8) multimode and (4) single mode fiber optic strands will be distributed to the first floor communications closet (located in the same room as the building entrance), (24) multimode and (4) single mode fiber optic strands will be distributed to the second floor communications closet and (24) multimode and (4) single mode fiber optic strands will be distributed to the third floor communications closet. All fiber optic distribution cabling will be installed in innerduct.

All fiber optic strands in the first, second and third floor communications closets will be terminated on ST type connectors and installed into ST type fiber optic patch panels. The fiber optic patch panels will be mounted into 19" free standing racks.

Copper backbone (tie) cabling will consist of (1) 25 pair category 5 cable installed in a loop between each closet. (1) 25 pair category 5 cable will run from the first floor communications closet to the second floor closet, (1) 25 pair category 5 cable will run from the second floor communications closet to the third floor closet and (1) 25 pair category 5 cable will run from the third floor communications closet to the first floor closet. All 25 pair category 5 backbone cables will terminate on 24 port category 5 patch panels which will be mounted in 19" free standing racks.

Reference drawing #'s Sample-02, Sample-05, Sample-06, Sample-07, Sample-08, Sample-11, Sample-12

Building 2 - Fiber optic backbone cabling for this facility will originate from the fiber optic distribution patch panels located in the first floor computer room. From this position (24) multimode and (6) single mode fiber optic strands will be distributed to the first floor communications closet and (24) multimode and (6) single mode fiber optic strands will be distributed to the second floor communications closet. In addition, (48) multimode and (12) single mode fiber optic strands which feed Building 1 and (48) multimode and (12) single mode fiber optic strands which feed Building 3 will terminate in the first floor computer room. All fiber optic distribution cabling will be installed in innerduct.

All fiber optic strands in the computer room and the first and second floor communications closets will be terminated on ST type connectors and installed into ST type fiber optic patch panels. The fiber optic patch panels will be mounted into 19" free standing racks.

Copper backbone (tie) cabling will consist of (1) 25 pair category 5 cable installed in a loop between each closet and the computer room. (1) 25 pair category 5 cable will run from the computer room to the first floor communications closet, (1) 25 pair category 5 cable will run from the first floor closet to the second floor communications closet and (1) 25 pair category 5 cable will run from the second floor closet to the computer room. All 25 pair category 5 backbone cables will terminate on 24 port category 5 patch panels which will be mounted in 19" free standing racks.

Reference drawing #'s Sample-03, Sample-05, Sample-09, Sample-10

Building 3 - Fiber optic backbone cabling for this facility will originate from the fiber distribution patch panel located in the building entrance area. From this position (24) multimode and (6) single mode fiber optic strands will be distributed to the first floor communications closet and (24) multimode and (6) single mode fiber optic strands will be distributed to the second floor communications closet. All fiber optic distribution cabling will be installed in innerduct.

All fiber optic strands in the building entrance area and the first and second floor communications closets will be terminated on ST type connectors and installed into ST type fiber optic patch panels. The fiber optic patch panels will be mounted into 19" free standing racks.

Copper backbone (tie) cabling will consist of (1) 25 pair category 5 cable installed between each closet. (1) 25 pair category 5 cable will run from the first floor communications closet to the second floor closet.

All 25 pair category 5 backbone cables will terminate on 24 port category 5 patch panels which will be mounted in 19" free standing racks.

Reference drawing #'s Sample-04, Sample-05, Sample-08, Sample-09

2.4 Outside Plant Cabling

Copper and fiber optic distribution cabling will be installed from Building 2 to Building 1 and Building 3. Cabling will be installed via the 4" conduits in the underground links. All fiber optic distribution cabling will be installed in innerduct.

Link 1 - Distribution cabling between Building 2 and Building 1 will require (1600) pairs of copper cable for voice, (48) multimode and (12) single mode fiber strands for data.

Voice distribution cabling will emanate from the building entrance area of Building 2 (to be defined) and will terminate at the building entrance room (first floor closet) of Building 1. Each end of the voice distribution cabling will terminate on 110 style punch blocks provided with the lightning protection systems.

Fiber optic distribution cabling will emanate from the computer room of Building 2 and will terminate at the building entrance room (first floor closet) of Building 1. Fiber optic cabling will terminate on ST type connectors and will be installed into ST type fiber optic patch panels. Fiber optic patch panels will be 19" rack mounted at each end. All fiber optic distribution cabling will be installed in innerduct.

Link 2 - Distribution cabling between Building 2 and Building 3 will require (1200) pairs of copper cable for voice, (48) multimode and (12) single mode fiber strands for data.

Voice distribution cabling will emanate from the building entrance area of Building 2 (to be defined) and will terminate at the building entrance area of Building 3 (to be defined). Each end of the voice distribution cabling will terminate on 110 style punch blocks provided with the lightning protection systems.

Fiber optic distribution cabling will emanate from the computer room of Building 2 and will terminate at the building entrance area of Building 3. Fiber optic cabling will terminate on ST type connectors and will be installed into ST type fiber optic patch panels. Fiber optic patch panels will be 19" rack mounted at Building 2 and wall mounted at Building 3. All fiber optic distribution cabling will be installed in innerduct.

2.5 Lightning Protection System

A lightning protection system will be installed at the building entrance area of each facility. All outside plant voice distribution cabling will terminate on these systems. The lightning protector panels will be mounted on plywood backboards. Plywood backboards are to be supplied and installed (where required) under this contract.

The lightning protection system will be connected to a suitable earth ground in each of the (4) building entrance areas.

3.0 Approved Manufacturers

3.1 Copper and Fiber Optic Cable

AT&T Network Systems
505 North 51st Avenue
Phoenix, AZ 85043
(800) 344-0223

Berk-Tek
132 White Oak Rd.
New Holland, PA 17557
(717) 354-6200

CommScope, Inc.
99 Brimstone Corner Rd.
Hancok, New Hampshire 03449
(800) 422-9961

General Cable Corp.
160 Fieldcrest Ave.
Edison, NJ 08837
(800) 526-4391

Mohawk Wire & Cable Corp.
9 Mohawk Dr.
Leominster, MA 01453
(800) 422-9961

3.2 Termination Hardware

AT&T Network Systems
505 North 51st Avenue
Phoenix, AZ 85043
(800) 344-0223

Ortronics
595 Greenhaven Rd.
Pawcatuck, CT 06379
(202) 599-1760

The Siemon Company
76 Westbury Park Rd.
Waterbury, CT 06795
(203) 275-2523

Thomas & Betts Corp.
1555 Lynnfield Rd.
Memphis, TN 38119
(901) 682-8221

3M Telecom Systems Group
6801 River Place Blvd.
Austin, TX 78726-9000
(800) 426-8688

Mod-Tap
P.O. Box 706
Harvard, MA 01451
(508) 772-4884

3.3 Lightning Protection

AT&T Network Systems
505 North 51st Avenue
Phoenix, AZ 85043
(800) 344-0223

Porta Systems Corp.
575 Underhill Blvd.
Syosset, NY 11791
(800) 937-6782

3.4 Miscellaneous Hardware

Caddy - Erico Fastening Products
34600 Solon Rd.
Solon, OH 44139-2695
(800) 252-2339

Hevi-duty / Nelson Firestop
4041 South Sheridan Rd.
Tulsa, OK 74145
(918) 627-5530

Homaco, Inc.
1875 West Fullerton avenue
Chicago, IL 60614
(312) 384-5575

Newton
111 East 'A' St. Box 727
Butner, NC 27509
(919) 575-6426

Ortronics
595 Greenhaven Rd.
Pawcatuck, CT 06379
(202) 599-1760

Pyramid Industries Inc.
1422 Irwin Dr.
P.O. Box 8439
Erie, PA 16505-0439
(814) 455-7587

Chatsworth Products Inc.
9541 Mason Ave.
Chatsworth, CA 91311
(818) 882-8595

Panduit / Network Systems Division
16530 163rd Street
Lockport, IL 60441
(800) 777-3300

4.0 Product Specifications

4.1 Copper and Fiber Optic Cable

4.1.1 General Cable Part # 7525868 - 600 Pair / 24 AWG Filled Foam Skin Qualpeth Cable used for voice distribution cabling between the three XYZ Corporation facilities. This cable will be installed within the 4" conduits provided in the conduit links. Termination of this cable will be on the lightning protection system.

4.1.2 General Cable Part # 7525850 - 400 Pair / 24 AWG Filled Foam Skin Qualpeth Cable used for voice distribution cabling between the three XYZ Corporation facilities. This cable will be installed within the 4" conduits provided in the conduit links. Termination of this cable will be on the lightning protection system.

4.1.3 Berk-Tek Part # PDR12B048CB3510/15 - 48 Strand tight buffered FDDI specification indoor / outdoor multimode fiber optic cable used for fiber optic distribution cabling between the three XYZ Corporation facilities. This cable will be installed within the 4" conduits provided in the conduit links. Termination will be on ST connectors which will be installed into ST to ST fiber optic patch panels.

4.1.4 Berk-Tek Part # PDR012AB1515 - 12 Strand tight buffered indoor / outdoor single mode fiber optic cable used for fiber optic distribution cabling between the three XYZ Corporation facilities. This cable will be installed within the 4" conduits provided in the enclosed walkways. Termination will be on ST connectors which will be installed into ST to ST fiber optic patch panels.

4.1.5 Mohawk Part # M56128 - 100 Pair category 3 plenum rated cable used for intrabuilding voice distribution cabling. This cabling will terminate on 300 pair 110 style punch blocks.

4.1.6 Berk-Tek Part # 230514 - 25 Pair category 5 plenum rated cable used for data tie cabling to be installed between the communications closets. This cabling will terminate on 24 port patch panels.

4.1.7 Berk-Tek Part # PDP6B024CB3510/15 - 24 Strand tight buffered FDDI specification multimode fiber optic cable used for intrabuilding fiber optic distribution cabling. Termination will be on ST connectors which will be installed into ST to ST fiber optic patch panels.

4.1.8 Berk-Tek Part # PDP008CB3510/15 - 8 Strand tight buffered FDDI specification multimode fiber optic cable used for intrabuilding fiber optic distribution cabling. Termination will be on ST connectors which will be installed into ST to ST fiber optic patch panels.

4.1.9 Berk-Tek Part # PDP006AB1515 - 6 Strand tight buffered single mode fiber optic cable used for intrabuilding fiber optic distribution cabling. Termination will be on ST connectors which will be installed into ST to ST fiber optic patch panels.

4.1.10 Berk-Tek Part # ICP004AB0707 - 4 Strand tight buffered single mode fiber optic cable used for intrabuilding fiber optic distribution cabling. Termination will be on ST connectors which will be installed into ST to ST fiber optic patch panels.

4.1.11 Mohawk Part # M55988 - 4 pair UTP enhanced category 5 plenum cable used for workstation data cabling. This cable will be installed between the communications closet and each workstation. Termination at the workstation will be on dual RJ45 station outlets and 48 port patch panels in the closets.

4.1.12 Mohawk Part # M55760 - 4 pair UTP category 3 plenum cable used for workstation voice cabling. This cable will be installed between the communications closet and each workstation. Termination at the workstation will be on dual

RJ45 station outlets and 300 pair 110 style punch blocks in the closets.

4.2 Termination Hardware

4.2.1 Ortronics Part # OR40300158 - Single gang 3 unit white faceplate used for mounting of RJ45 modules.

4.2.2 Ortronics Part # OR60950009 - Dual RJ45 category 5 connector module used for termination of category 5 workstation data cables.

4.2.3 Ortronics Part # OR60940009 - Dual RJ45 category 4 connector module used for termination of category 3 workstation voice cables.

4.2.4 Ortronics Part # OR40300164 - Blank module used to fill the unused unit space in the single gang faceplate.

4.2.5 Ortronics Part # OR40300061 - Single gang low profile surface box used as workstation box for the modular furniture areas.

4.2.6 Ortronics Part # OR851004912 - 48 Port category 5 patch panel used in communications closets for termination of category 5 workstation data cables.

4.2.7 Ortronics Part # OR851004904 - 24 Port category 5 patch panel used in communications closets for termination of 25 pair tie cables.

4.2.8 Ortronics Part # OR615MMC-24P - Fiber Optic patch panel cabinet with drawer faceplates, 2 blanks, and supports up to 36 multimode or single mode fiber optic strands. Used in communications closets for termination point of intrabuilding fiber optic cabling.

4.2.9 Ortronics Part # OR615MMC-72P - Fiber Optic patch panel cabinet with drawer faceplates and supports up to 72 multimode or single mode fiber optic strands. Used as termination point of interbuilding fiber optic cabling.

4.2.10 Ortronics Part # OR615SMFC-24P - Surface Mount Fiber Optic Cabinet supports up to 24 multimide or single mode fiber optic strands. Used to patch inter building fiber optic distribution cabling.

4.2.11 Ortronics Part # OR615STMM6 - 6 ST multimode coupler panel used in fiber optic patch panel cabinets as connection point for terminated multimode fiber optic strands.

4.2.12 Ortronics Part # OR615STSM6 - 6 ST single mode coupler panel used in fiber optic patch panel cabinets as connection point for terminated single mode fiber optic strands.

4.2.13 3M Part # 6105 - Multimode field mountable ST connector used for termination of all multimode fiber optic strands.

4.2.14 3M Part # 8105 - single mode field mountable ST connector used for termination of all single mode fiber optic strands.

4.2.12 AT&T Part # 107 059 917 - 300 Pair 110 punch block with legs used to terminate voice workstation and distribution cabling where wall mounting is required.

4.2.13 AT&T Part # 107 059 925 - 300 Pair 110 punch block without legs used to terminate voice workstation and distribution cabling where rack mounting is required.

4.2.14 AT&T Part # 103 801 247 - 4 pair connecting block used with 110 punch blocks to terminate workstation voice cables.

4.2.15 AT&T Part # 103 801 247 - 5 pair connecting block used with 110 punch blocks to terminate voice distribution cables.

4.3 Lightning Protection

4.3.1 AT&T Part # 106 086 762 - 188 Type 100 pair lightning protector panel used for termination of outside plant voice distribution cabling.

4.3.2 AT&T Part # 104 401 856 - 4B-EW Series gas tube protector unit that installs into the 188 style protector panel used for protection of voice circuits.

4.4 Miscellaneous hardware

4.4.1 Ortronics Part # OR604004600 - 19" Free standing rack 7' high used for mounting data and fiber optic patch panels, wire management panels and 110 blocks.

4.4.2 Mod Tap Part # 25.B035G - Wire management panel used for management of patch cables within the equipment racks.

4.4.3 Ortronics Part # OR808004759 - Wire management panel used for management of cross connect wire on rack mounted punch blocks.

4.4.4 Ortronics Part # OR808004868 - Vertical wire management bracket used for management of cross connect wire on rack mounted punch blocks.

4.4.5 Pyramid Industries Part # 4404-06FR - 1.5" Flame retardent innerduct used for installation in 4" conduits between facilities.

4.4.6 Pyramid Industries Part # PLM150T - 1.5" Plenum rated innerduct used to protect intra building fiber optic distribution cabling.

4.4.7 Chatsworth Part # 3001-3-700 - Mounting bars used to install 300 pair punch blocks onto 19" free standing racks.

4.4.8 Hevi duty / Neslon Firestop Part # AA446 - Firestop putty used to seal any penetrations of sleeves or conduits used for communications cabling.

4.4.9 Newton - 12" Cable tray (and associated hardware) used to distribute cabling within the communications closets.

4.4.10 Panduit Part # F3X2LG6 - Duct used for cross connect management of wall mounted punch blocks.

4.5 Notes to Product Specifications

Only products specified in this section or their approved equivalents are to be quoted under this RFP. Prior to the start of the project, product specification submittals will be required of all quoted products.

5.0 Requirements

5.1 Installation

The contractor shall provide and install all copper and fiber optic cable, connectors, patch panels, racks, punch blocks and associated hardware required to supply a complete cable plant, as defined in this RFP, to the XYZ Corporation.

All cable, connectors, patch panels, punch blocks and equipment racks shall be installed in accordance with manufacturers specifications.

Ladder rack will be required in the communications closets only. For the purposes of this bid assume 20 feet of ladder rack is to be installed per communications closet.

Where there is no ladder rack, workstation cabling must be supported by a suitable cable support product every 5 feet. All cable is to be neatly dressed into its termination point. In certain areas workstation cabling to modular furniture groups will be fed from the floor below.

Free standing equipment racks are to be securely mounted to the floor. Any sleeves or conduit penetrations must be firestopped subsequent to cable installation.

Workstation outlets will be installed into single gang surface boxes at modular furniture stations and into the drywall in hardwall offices. It is the contractors responsibility to provide and install a box eliminator in the offices where an electrical backbox does not exist.

5.2 Labeling

All workstation, copper and fiber distribution cables, workstation outlets, patch panels and punch blocks must be clearly labeled in accordance with the labeling specifications to be provided by the XYZ Corporation at a later date.

Only machine generated labels will be accepted.

5.3 Testing

All category 3 workstation cabling and all voice distribution cabling (inside and outside plant) will be continuity and polarity tested. Any defective category 3 workstation cabling shall be replaced by the contractor.

All category 5 workstation cable and category 5 data tie cable will be tested for Line Mapping (opens, shorts, reversals), attenuation, distance, near end and far end crosstalk (NEXT), mutual capacitance and signal to noise ratio (SNR). All tests must pass nominal EIA/TIA requirements for category 5 certification. Certification testing must be performed by a Wavetek LAN Pro XL or equivalent unit.

All multimode fiber optic strands (inside and outside plant) will be power meter tested at 850nm and 1300nm. Single mode fiber optic strands (inside and outside plant) will be power meter tested at 1300nm and 1550nm. Any strand having a loss greater than 2.0 db will be unacceptable. Any fiber optic strand testing over this limit will be repaired or replaced. All single mode and multimode fiber strands will be power meter link tested from each communications closet in Building 1 and Building 3 to the computer room in Building 2.

5.4 Documentation

As-built drawings are to be provided showing cable path and workstation outlet numbers.

All category 3 cabling tests are to be provided in a spreadsheet format showing test date, technician and test results.

Category 5 certification tests results are to be supplied in magnetic media in ASCII format that can be read by any standard word processor or spreadsheet.

5.5 Submittals

Manufacturers product specification sheets must be submitted for every product to be used in this cable plant within 1 week of project award.

5.6 Codes and Standards

All work shall conform to the applicable codes and standards of the agencies listed below. When a conflict between standards arises the more stringent standard is to be followed.

NEC - National Electric Code, Article 800 Communications Circuits

Local Electrical Codes and Ordinances

NFPA - National Fire Protection Association

NEMA - National Electrical Manufacturers Association

EIA/TIA - Electronic Industries Association / Telecommunications Industries Association, 568 Commercial Building Wiring Standards, 568 TSB 40 Twisted Pair Connecting Categories

ANSI - American National Standards Institute, X3T9.5 FDDI Standard

IEEE - Institute of Electrical and Electronic Engineers

FCC - Federal Communications Commission

UL - Underwriters Laboratories

OSHA - Occupational Health and Safety Administration

5.7 Warranty

Contractor shall supply in writing a (5) year warranty against defects in workmanship or materials on the installed cable plant. Any defects discovered during this (5) year period shall be repaired or replaced at no charge to the XYZ Corporation.

5.8 Changes and Alternates

Any changes or deviations from the cable plant specification or approved products and manufacturers defined in this document will be considered non-responsive by the XYZ Corporation.

Any changes or deviations from the pricing formats defined in the appendices of this document will be considered non-responsive by the XYZ Corporation.

5.9 Project Management

Contractor will provide a Project Manager who will act as a single point of contact between contractor and the XYZ Corporation. This project manager must attend construction meetings and have the authority to make decisions regarding project implementation and manpower commitments. Additionally, the chosen Project Manager must have a proven track record of managing installations of similar size.

Project Management personnel cannot be changed without prior notice and approval of the XYZ Corporation.

5.10 Right of Refusal

The XYZ Corporation reserves the right to accept or reject any or all responses to this RFP and / or enter into negotiations with more than one qualified supplier if it is in the best interest of the XYZ Corporation.

Any costs associated with responding to this RFP are the responsibility of the contractor.

5.11 Work by Others

Any conduit, stub ups, poke thru's or core bores which may be required for this project will be provided by others. It is the contractors responsibility to provide and install a box eliminator in the offices where an electrical backbox does not exist.

5.12 Insurance's

Bidders are to provide a certificate of insurance with their response outlining insurance coverage's which will be provided during the course of this project.

5.13 Debris Removal

It is the responsibility of the contractor to discard debris on a daily basis and to keep the work area clean.

5.14 Material Handling

Contractor is responsible for the handling of all materials and all costs associated with material handling. It is the contractors responsibility to coordinate delivery schedules with the project schedule.

5.15 Subcontractors

Contractor will notify XYZ Corporation of any subcontractors which may be used for this project and will define the scope of work to be performed by the subcontractor. If the subcontractor fails to perform the specified scope of work, the XYZ Corporation shall have the right to remove subcontractor from the project. In no way will this release the contractor from his obligations under the contract.

Contractor shall hold harmless and indemnify the XYZ Corporation against any claims, suits, liens or other actions made by subcontractor due to this agreement.

5.16 Compliance

Bidders shall provide with their response a compliance summary stating compliance or exception to every article of this RFP. Exception to any article must be accompanied by a full explanation.

6.0 Pricing

6.1 Pricing Format

Pricing for this project is to include all labor and all materials to install, tag and test all components of this cable plant as defined in this RFP. Pricing must be provided in the format as defined in Appendix 'B' of this RFP. Any deviation from this format will be considered non-responsive by the XYZ Corporation

6.2 Unit Prices

Unit prices to be provided for this project include unit cost for a standard station drop, cross connects and unit prices for various length category 5 patch cords and fiber optic jumpers. Pricing must be provided in the format as defined in Appendix 'C' of this RFP. Any deviation from this format will be considered non-responsive by the XYZ Corporation.

6.3 Labor Rates

Hourly labor rates are to be provided for "Straight Time", "Overtime" and "Double Time" with a definition of conditions which these rates apply (Appendix 'D').

6.4 Invoicing Format

A sample invoicing format is supplied in Appendix 'E' of this RFP. This format includes breakdowns showing current work completed and previous work completed. Deviation from this format will delay payment until the proper invoicing format is supplied.

7.0 Bid Schedule

Issue RFP	3/1/XX
Bid Conference	3/8/XX
Response Due Date	3/15/XX
Project Award Date	3/21/XX

Bid Conference will be held at Building 3, 3 Main St., Anytown USA starting at 10:00 AM. Respondents are to meet in the lobby of this facility.

Three (3) copies of bid responses are due by 3:00 PM on the above specified date. Bid responses are to be sent to "Your Name" as specified on page 1 of this RFP. There will be no extensions granted for the response due date.

8.0 Bidder Qualifications

8.1 References

Bidders must provide (3) references for project completed within the past two years of similar size. References must include Company name, address, phone number and contact person. Also include a brief summary of the actual projects performed for the customer.

8.2 Resumes

Provide resumes for several Project Managers, one of which who will be chosen to manage this project.

8.3 Organizational Structure

Provide a definition of current staffing. This should include number of Project Managers, number of installation technicians, number and position of management personnel etc..

8.4 Licensing

Bidder must provide a list of all applicable licenses currently held by installation personnel.

8.5 Manpower

Due to the implementation plan of this project, a large contingent of manpower may be required for limited durations. Provide a description of manpower availability and how this requirement will be met.

Appendix 'A'

Drawings

Campus Layout

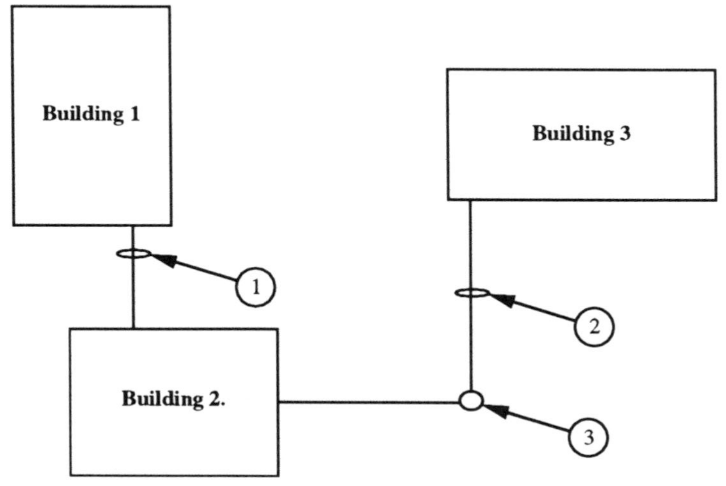

Building 1

Building 2.

Building 3

1

2

3

Notes:

1. (4) 4" Conduits.
2. (4) 4" Conduits
3. Manhole

D&D

Project - *XYZ Corporation*	
Drawing # - *Sample - 01*	*Rev -* *0*
Drawn By -	*Date -*

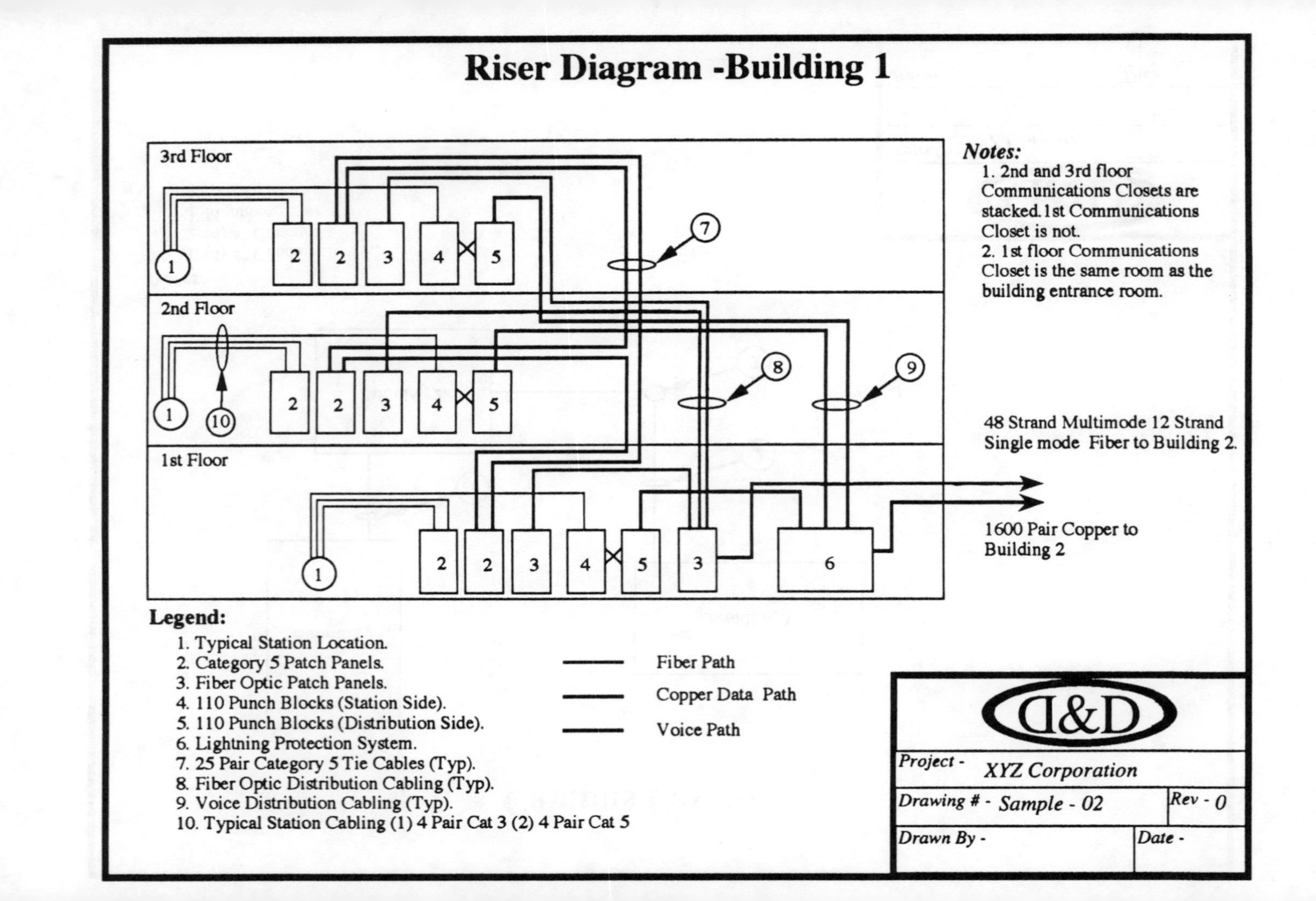
Riser Diagram -Building 1
3rd Floor
2nd Floor
1st Floor
1
2
2
3
4
5
7
8
9
10
6
Notes:
1. 2nd and 3rd floor Communications Closets are stacked. 1st Communications Closet is not.
2. 1st floor Communications Closet is the same room as the building entrance room.
48 Strand Multimode 12 Strand Single mode Fiber to Building 2.
1600 Pair Copper to Building 2
Legend:
1. Typical Station Location.
2. Category 5 Patch Panels.
3. Fiber Optic Patch Panels.
4. 110 Punch Blocks (Station Side).
5. 110 Punch Blocks (Distribution Side).
6. Lightning Protection System.
7. 25 Pair Category 5 Tie Cables (Typ).
8. Fiber Optic Distribution Cabling (Typ).
9. Voice Distribution Cabling (Typ).
10. Typical Station Cabling (1) 4 Pair Cat 3 (2) 4 Pair Cat 5
Fiber Path
Copper Data Path
Voice Path
D&D
Project - XYZ Corporation
Drawing # - Sample - 02
Rev - 0
Drawn By -
Date -

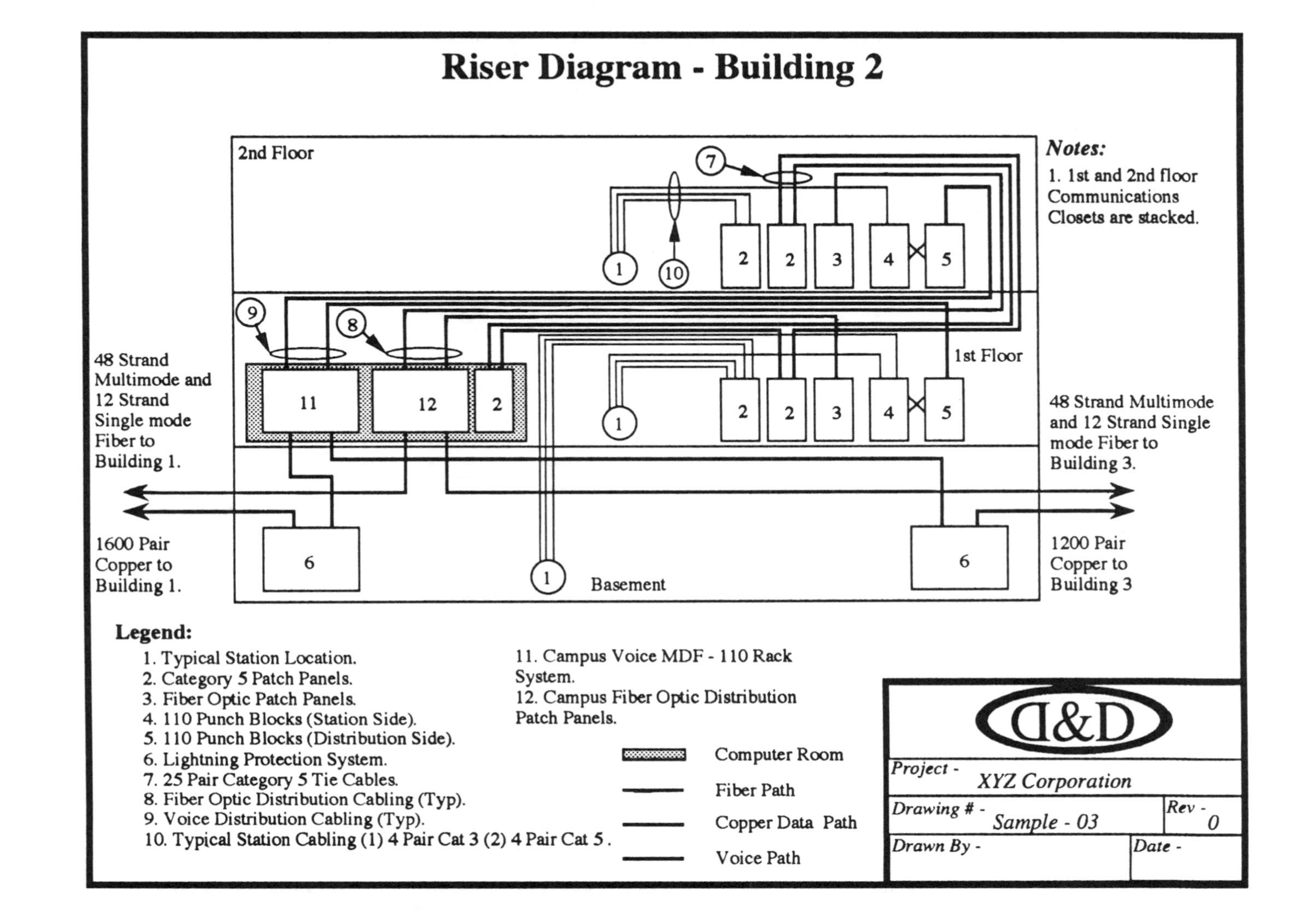
Riser Diagram - Building 2
2nd Floor
1st Floor
Basement
Notes:
1. 1st and 2nd floor Communications Closets are stacked.
48 Strand Multimode and 12 Strand Single mode Fiber to Building 1.
1600 Pair Copper to Building 1.
48 Strand Multimode and 12 Strand Single mode Fiber to Building 3.
1200 Pair Copper to Building 3
Legend:
1. Typical Station Location.
2. Category 5 Patch Panels.
3. Fiber Optic Patch Panels.
4. 110 Punch Blocks (Station Side).
5. 110 Punch Blocks (Distribution Side).
6. Lightning Protection System.
7. 25 Pair Category 5 Tie Cables.
8. Fiber Optic Distribution Cabling (Typ).
9. Voice Distribution Cabling (Typ).
10. Typical Station Cabling (1) 4 Pair Cat 3 (2) 4 Pair Cat 5 .
11. Campus Voice MDF - 110 Rack System.
12. Campus Fiber Optic Distribution Patch Panels.
Computer Room
Fiber Path
Copper Data Path
Voice Path
D&D
Project - XYZ Corporation
Drawing # - Sample - 03
Rev - 0
Drawn By -
Date -

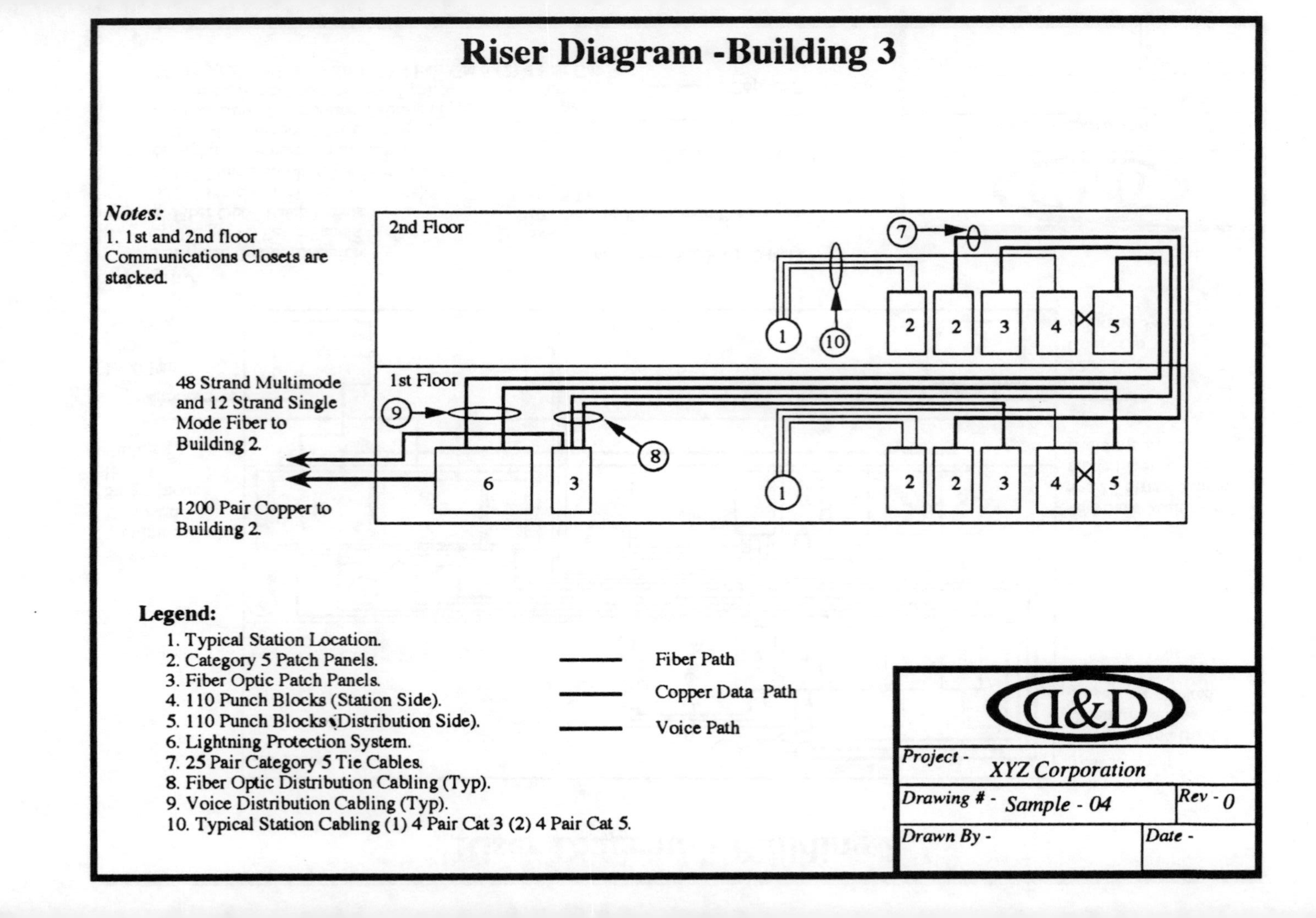
Riser Diagram -Building 3
Notes:
1. 1st and 2nd floor Communications Closets are stacked.
2nd Floor
1st Floor
48 Strand Multimode and 12 Strand Single Mode Fiber to Building 2.
1200 Pair Copper to Building 2.
1
2
3
4
5
6
7
8
9
10
Legend:
1. Typical Station Location.
2. Category 5 Patch Panels.
3. Fiber Optic Patch Panels.
4. 110 Punch Blocks (Station Side).
5. 110 Punch Blocks (Distribution Side).
6. Lightning Protection System.
7. 25 Pair Category 5 Tie Cables.
8. Fiber Optic Distribution Cabling (Typ).
9. Voice Distribution Cabling (Typ).
10. Typical Station Cabling (1) 4 Pair Cat 3 (2) 4 Pair Cat 5.
Fiber Path
Copper Data Path
Voice Path
D&D
Project - XYZ Corporation
Drawing # - Sample - 04
Rev - 0
Drawn By -
Date -

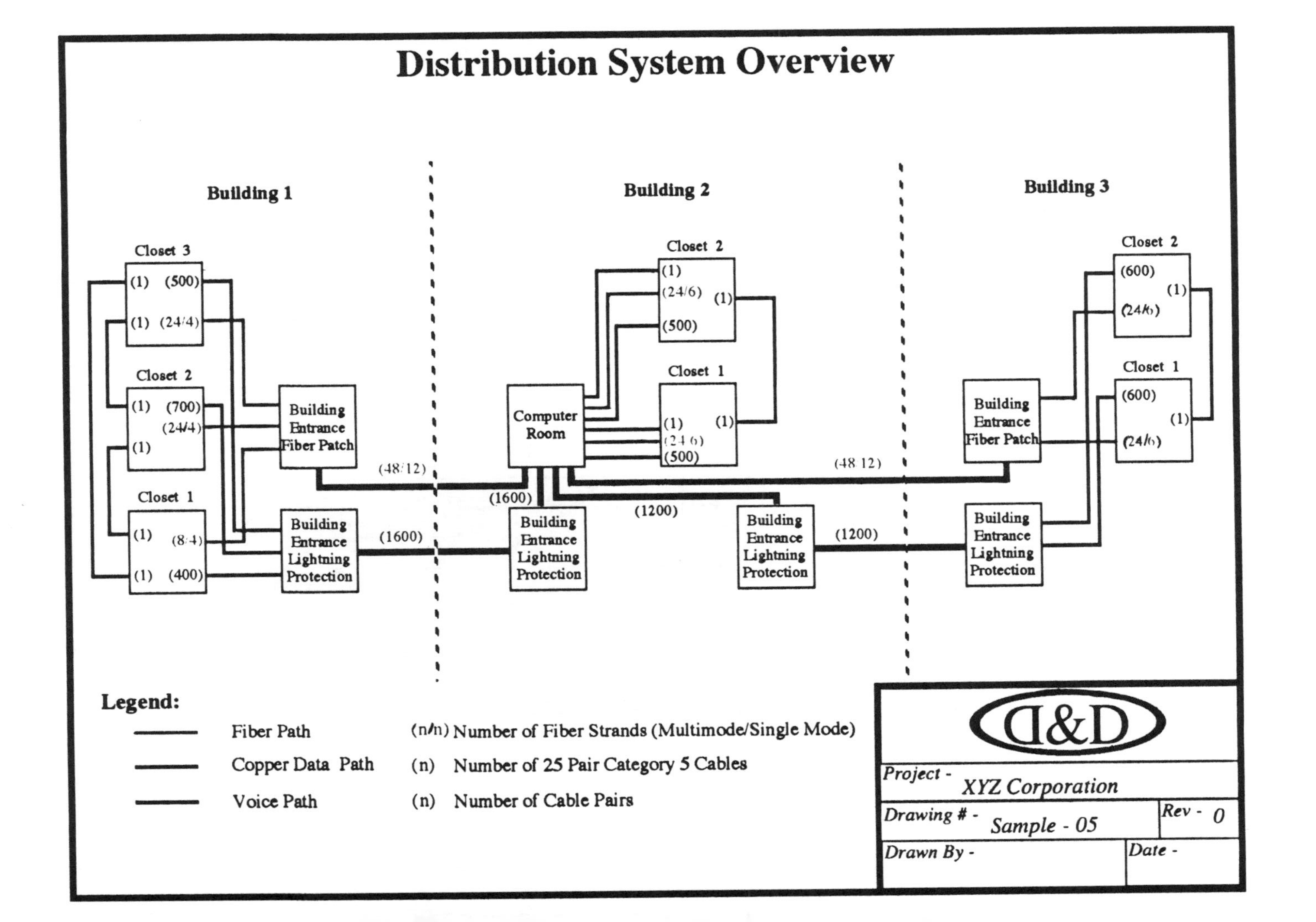
Distribution System Overview
Building 1
Building 2
Building 3
Closet 3
(1) (500)
(1) (24/4)
Closet 2
(1) (700)
(24/4)
(1)
Closet 1
(1) (8/4)
(1) (400)
Building Entrance Fiber Patch
Building Entrance Lightning Protection
(48/12)
(1600)
Computer Room
Closet 2
(1)
(24/6)
(1)
(500)
Closet 1
(1)
(1)
(24/6)
(500)
(1600)
(1200)
Building Entrance Lightning Protection
Building Entrance Lightning Protection
(48/12)
(1200)
Closet 2
(600)
(1)
(24/6)
Closet 1
(600)
(1)
(24/6)
Building Entrance Fiber Patch
Building Entrance Lightning Protection
Legend:
Fiber Path
Copper Data Path
Voice Path
(n/n) Number of Fiber Strands (Multimode/Single Mode)
(n) Number of 25 Pair Category 5 Cables
(n) Number of Cable Pairs
D&D
Project - XYZ Corporation
Drawing # - Sample - 05
Rev - 0
Drawn By -
Date -

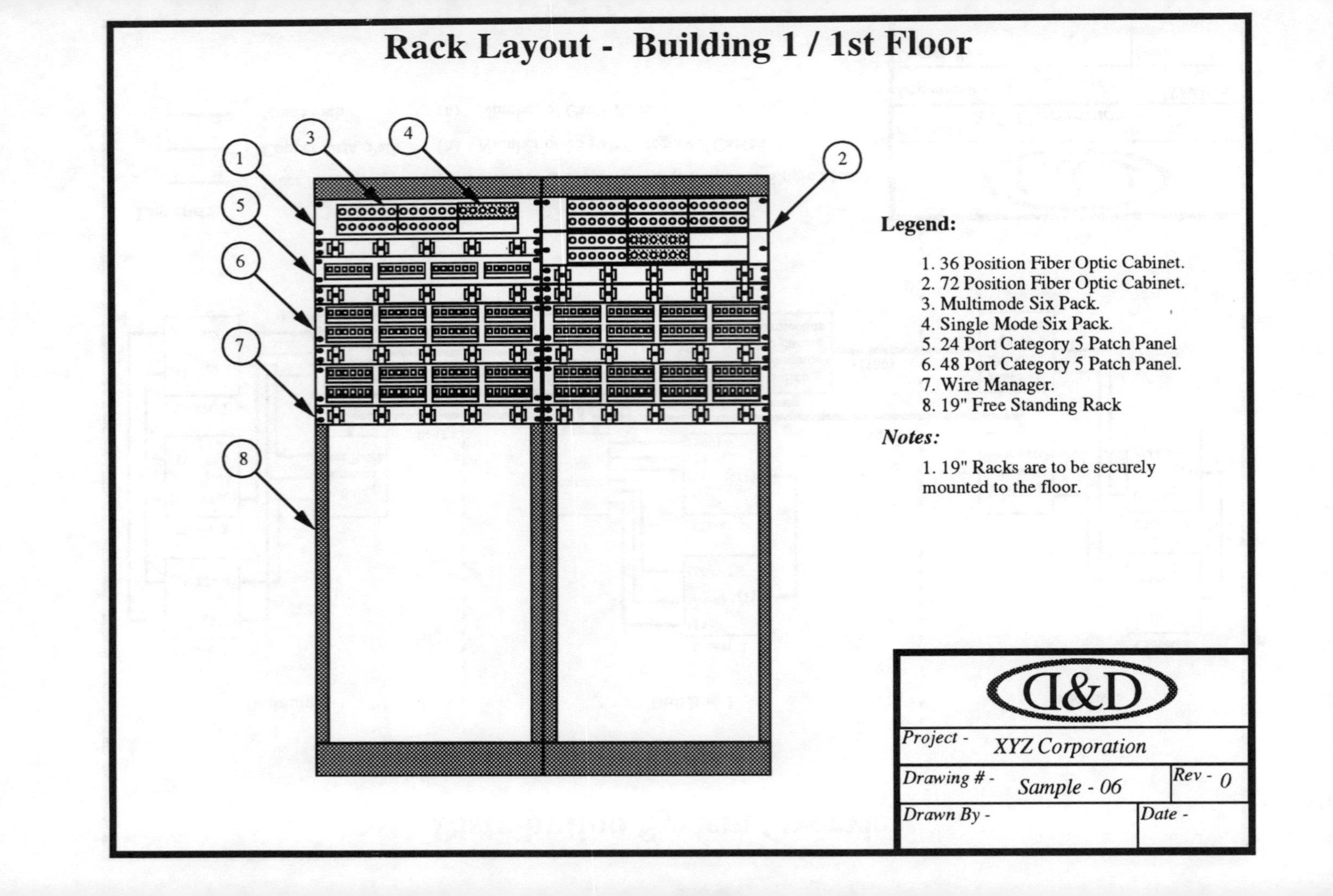
Rack Layout - Building 1 / 1st Floor
1
2
3
4
5
6
7
8
Legend:
1. 36 Position Fiber Optic Cabinet.
2. 72 Position Fiber Optic Cabinet.
3. Multimode Six Pack.
4. Single Mode Six Pack.
5. 24 Port Category 5 Patch Panel
6. 48 Port Category 5 Patch Panel.
7. Wire Manager.
8. 19" Free Standing Rack
Notes:
1. 19" Racks are to be securely mounted to the floor.
D&D
Project - XYZ Corporation
Drawing # - Sample - 06
Rev - 0
Drawn By -
Date -

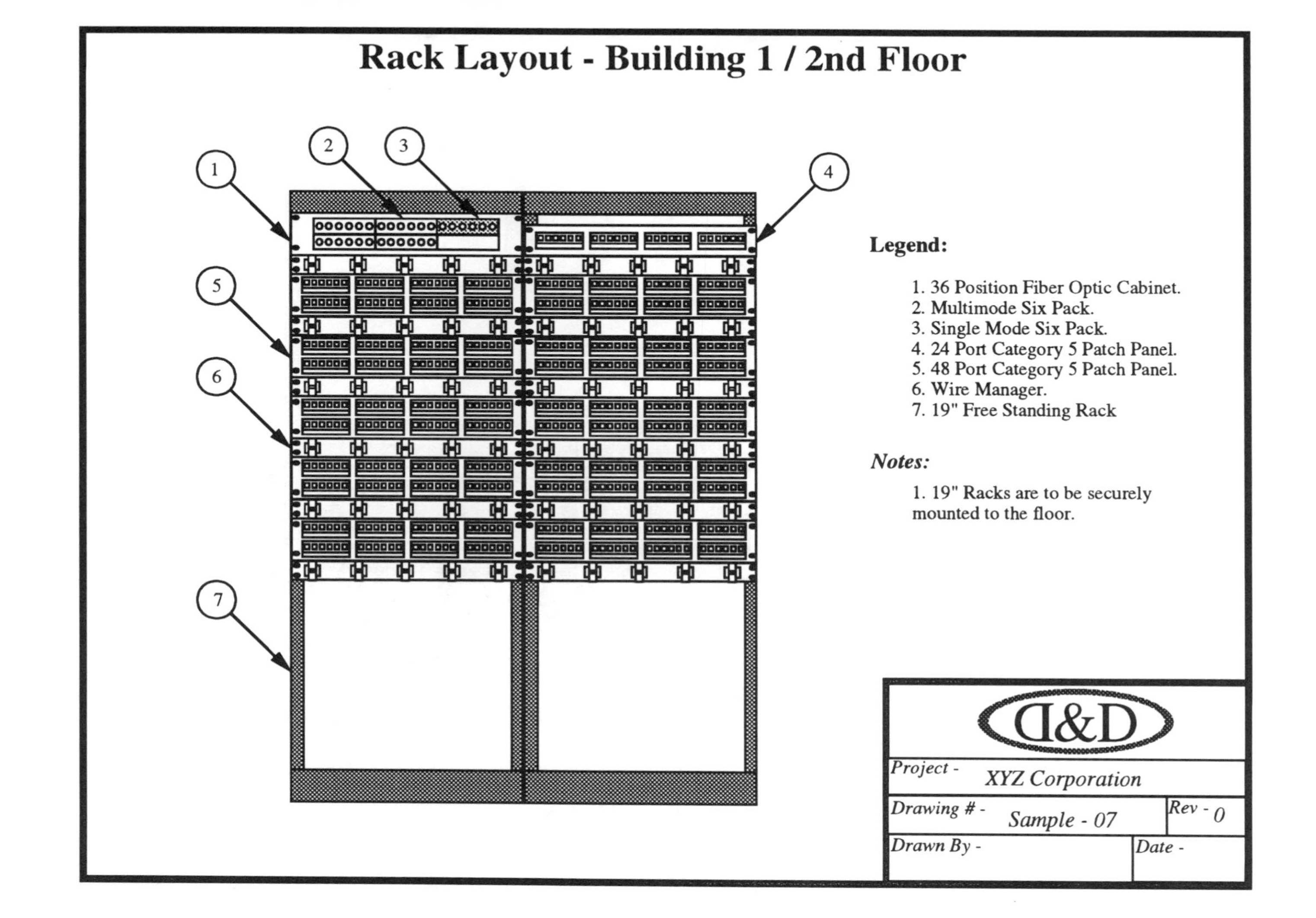
Rack Layout - Building 1 / 2nd Floor
1
2
3
4
5
6
7
Legend:
1. 36 Position Fiber Optic Cabinet.
2. Multimode Six Pack.
3. Single Mode Six Pack.
4. 24 Port Category 5 Patch Panel.
5. 48 Port Category 5 Patch Panel.
6. Wire Manager.
7. 19" Free Standing Rack
Notes:
1. 19" Racks are to be securely mounted to the floor.
D&D
Project - XYZ Corporation
Drawing # - Sample - 07
Rev - 0
Drawn By -
Date -

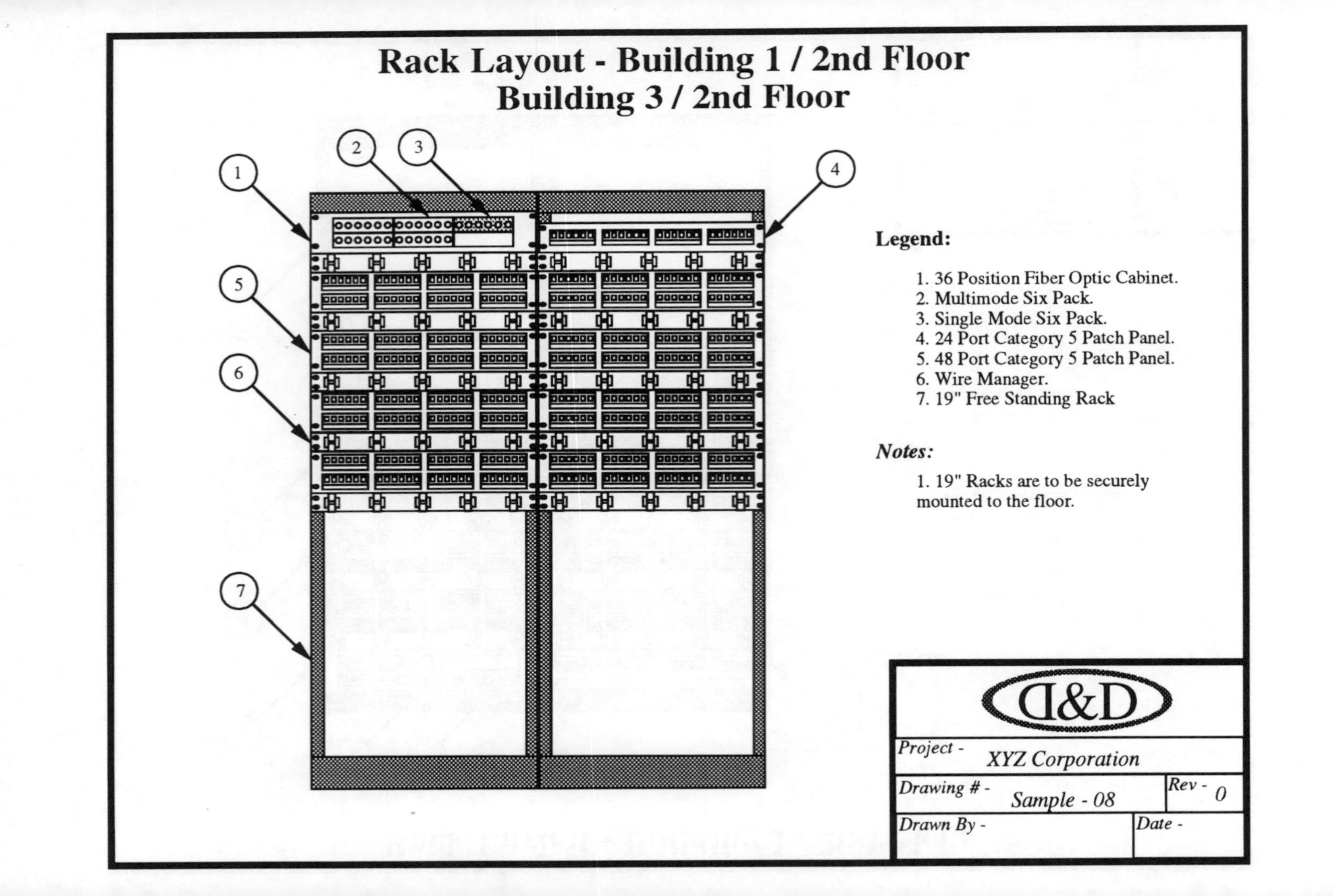
Rack Layout - Building 1 / 2nd Floor
Building 3 / 2nd Floor
1
2
3
4
5
6
7
Legend:
1. 36 Position Fiber Optic Cabinet.
2. Multimode Six Pack.
3. Single Mode Six Pack.
4. 24 Port Category 5 Patch Panel.
5. 48 Port Category 5 Patch Panel.
6. Wire Manager.
7. 19" Free Standing Rack
Notes:
1. 19" Racks are to be securely mounted to the floor.
D&D
Project - XYZ Corporation
Drawing # - Sample - 08
Rev - 0
Drawn By -
Date -

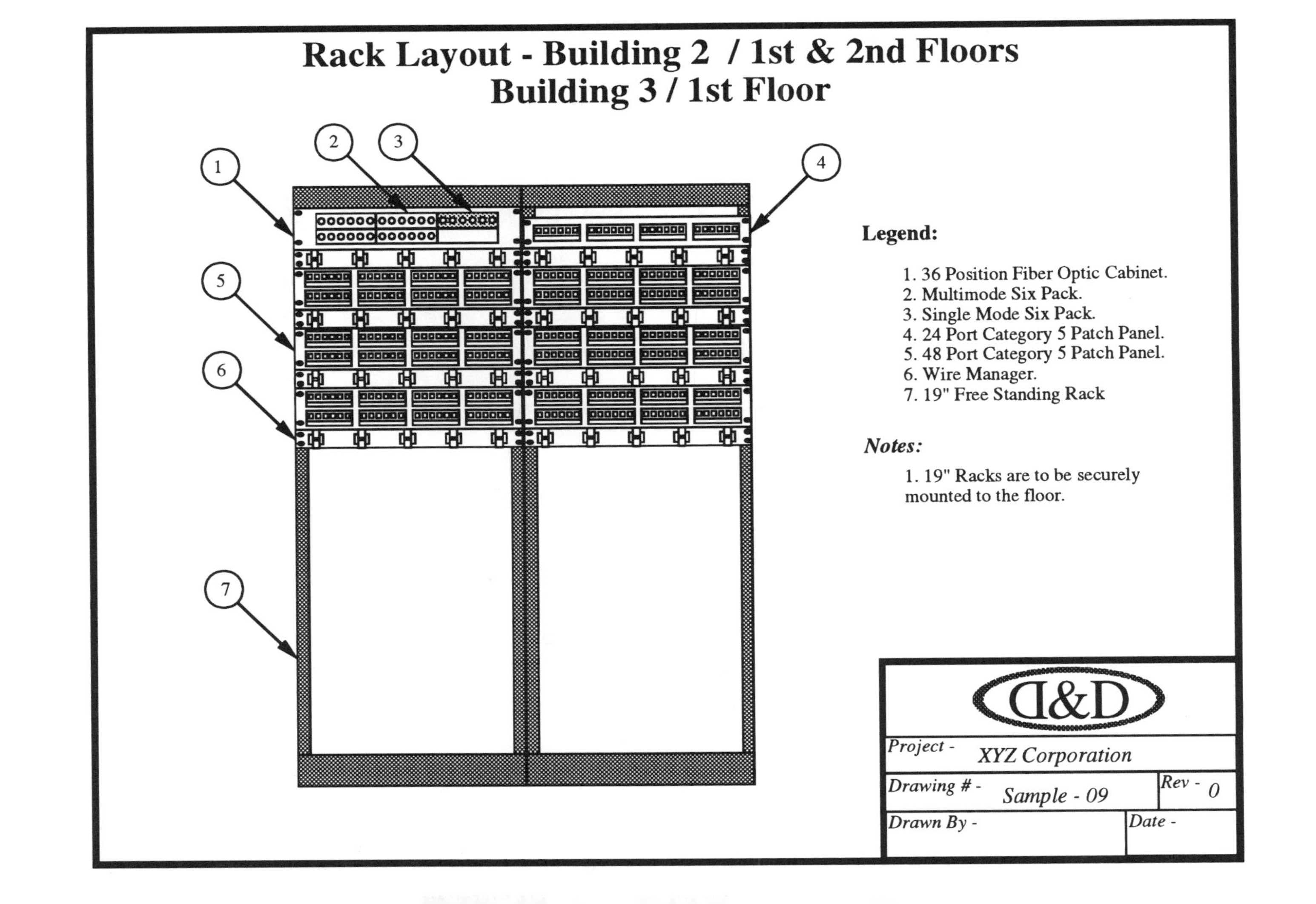
Rack Layout - Building 2 / 1st & 2nd Floors
Building 3 / 1st Floor
1
2
3
4
5
6
7
Legend:
1. 36 Position Fiber Optic Cabinet.
2. Multimode Six Pack.
3. Single Mode Six Pack.
4. 24 Port Category 5 Patch Panel.
5. 48 Port Category 5 Patch Panel.
6. Wire Manager.
7. 19" Free Standing Rack
Notes:
1. 19" Racks are to be securely mounted to the floor.
D&D
Project - XYZ Corporation
Drawing # - Sample - 09
Rev - 0
Drawn By -
Date -

Rack Layout - Building 2 / Computer Room

1
2
3
4
5
6
7

Legend:

1. 72 Position Fiber Optic Cabinet.
2. Multimode Six Pack.
3.Single Mode Six Pack.
4. 36 Position Fiber Optic Cabinet.
5. Wire Manager
6. 24 Port Category 5 Patch Panel
7. 19" Free Standing Rack

Notes:

1. 19" Racks are to be securely mounted to the floor.

D&D

Project - XYZ Corporation	
Drawing # - Sample - 10	*Rev -* 0
Drawn By -	*Date -*

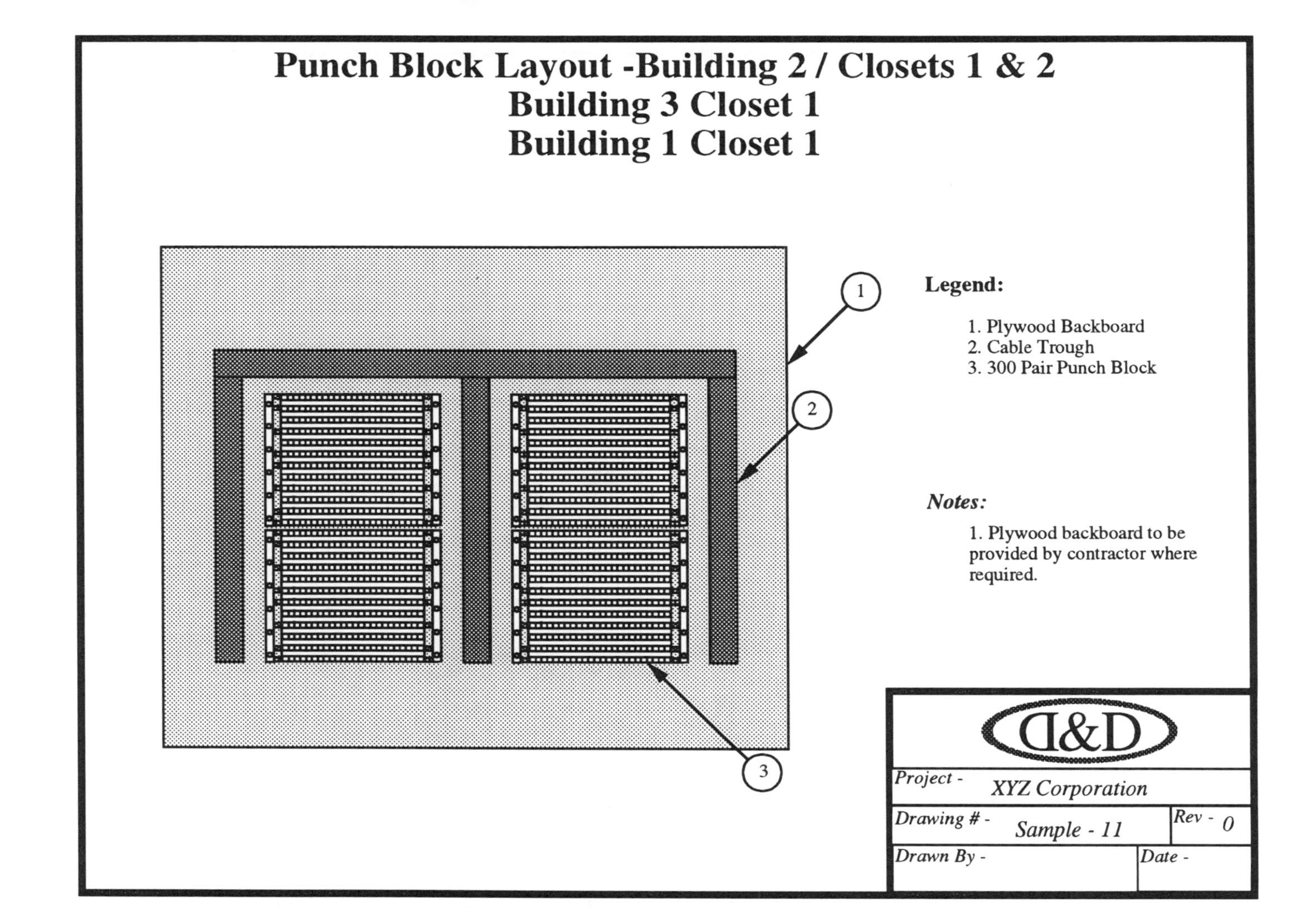
Punch Block Layout -Building 2 / Closets 1 & 2
Building 3 Closet 1
Building 1 Closet 1
1
2
3
Legend:
1. Plywood Backboard
2. Cable Trough
3. 300 Pair Punch Block
Notes:
1. Plywood backboard to be provided by contractor where required.
D&D
Project - XYZ Corporation
Drawing # - Sample - 11
Rev - 0
Drawn By -
Date -

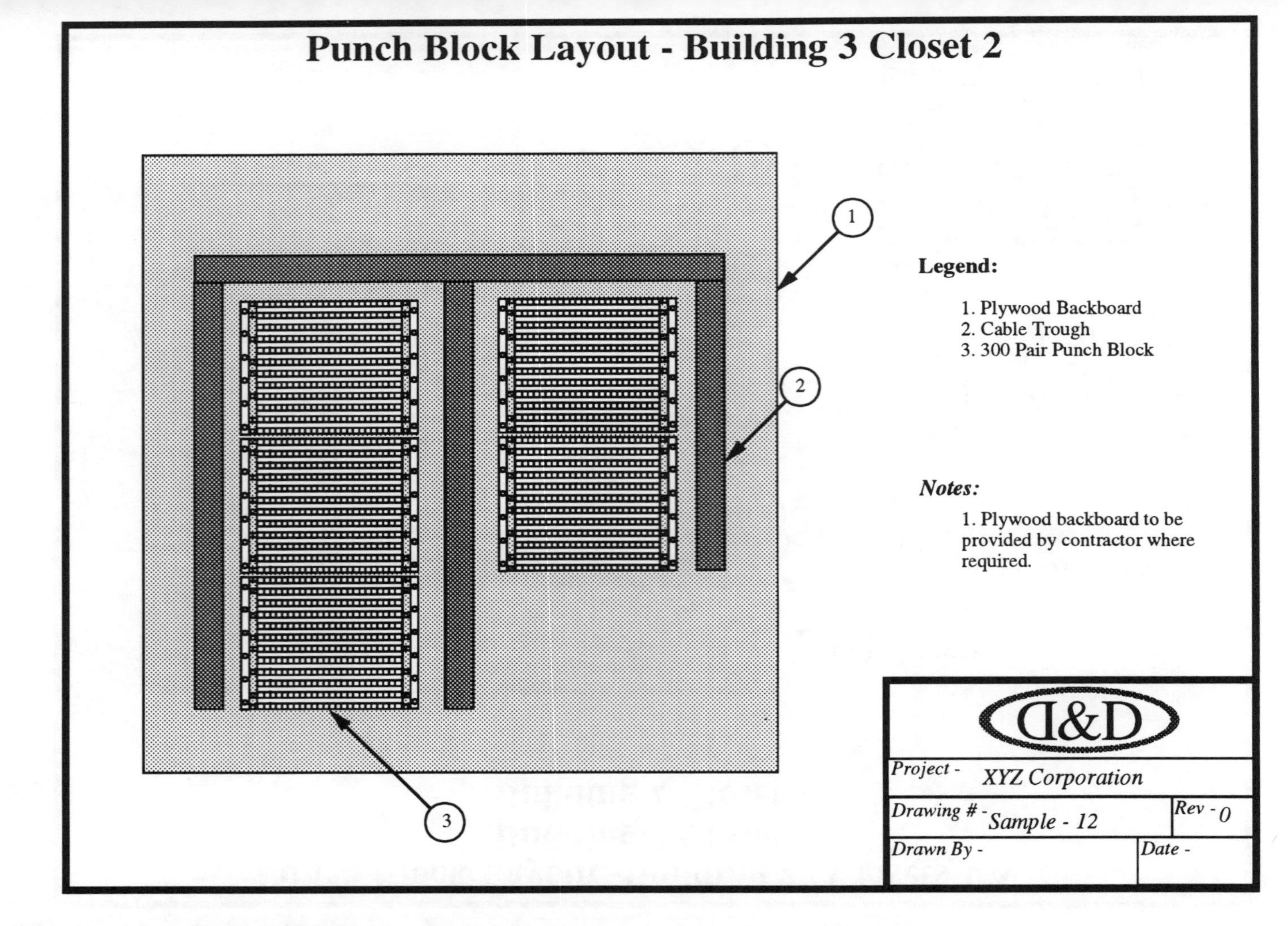
Punch Block Layout - Building 3 Closet 2
1
2
3
Legend:
1. Plywood Backboard
2. Cable Trough
3. 300 Pair Punch Block
Notes:
1. Plywood backboard to be provided by contractor where required.
D&D
Project - XYZ Corporation
Drawing # - Sample - 12
Rev - 0
Drawn By -
Date -

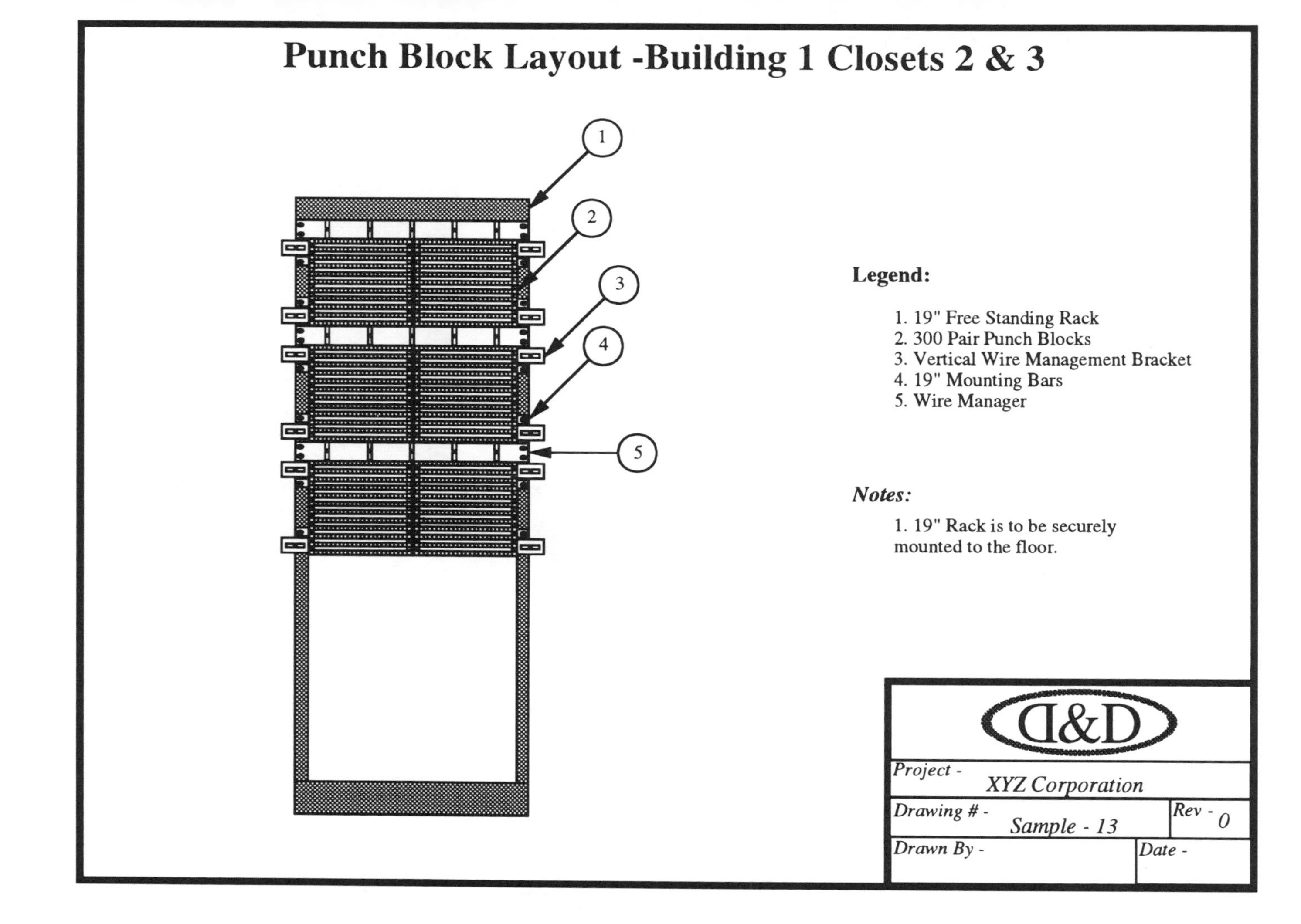
Punch Block Layout -Building 1 Closets 2 & 3
1
2
3
4
5
Legend:
1. 19" Free Standing Rack
2. 300 Pair Punch Blocks
3. Vertical Wire Management Bracket
4. 19" Mounting Bars
5. Wire Manager
Notes:
1. 19" Rack is to be securely
mounted to the floor.
D&D
Project - XYZ Corporation
Drawing # - Sample - 13
Rev - 0
Drawn By -
Date -

Punch Block Layout - Building 2 / MDF (Distribution Side)

Legend:

1. 19" Free Standing Rack
2. 300 Pair Punch Blocks
3. Vertical Wire Management Bracket
4. 19" Mounting Bars
5. Wire Manager

Notes:

1. 19" Rack is to be securely mounted to the floor.

D&D

Project - XYZ Corporation	
Drawing # - Sample - 14	*Rev -* 0
Drawn By -	*Date -*

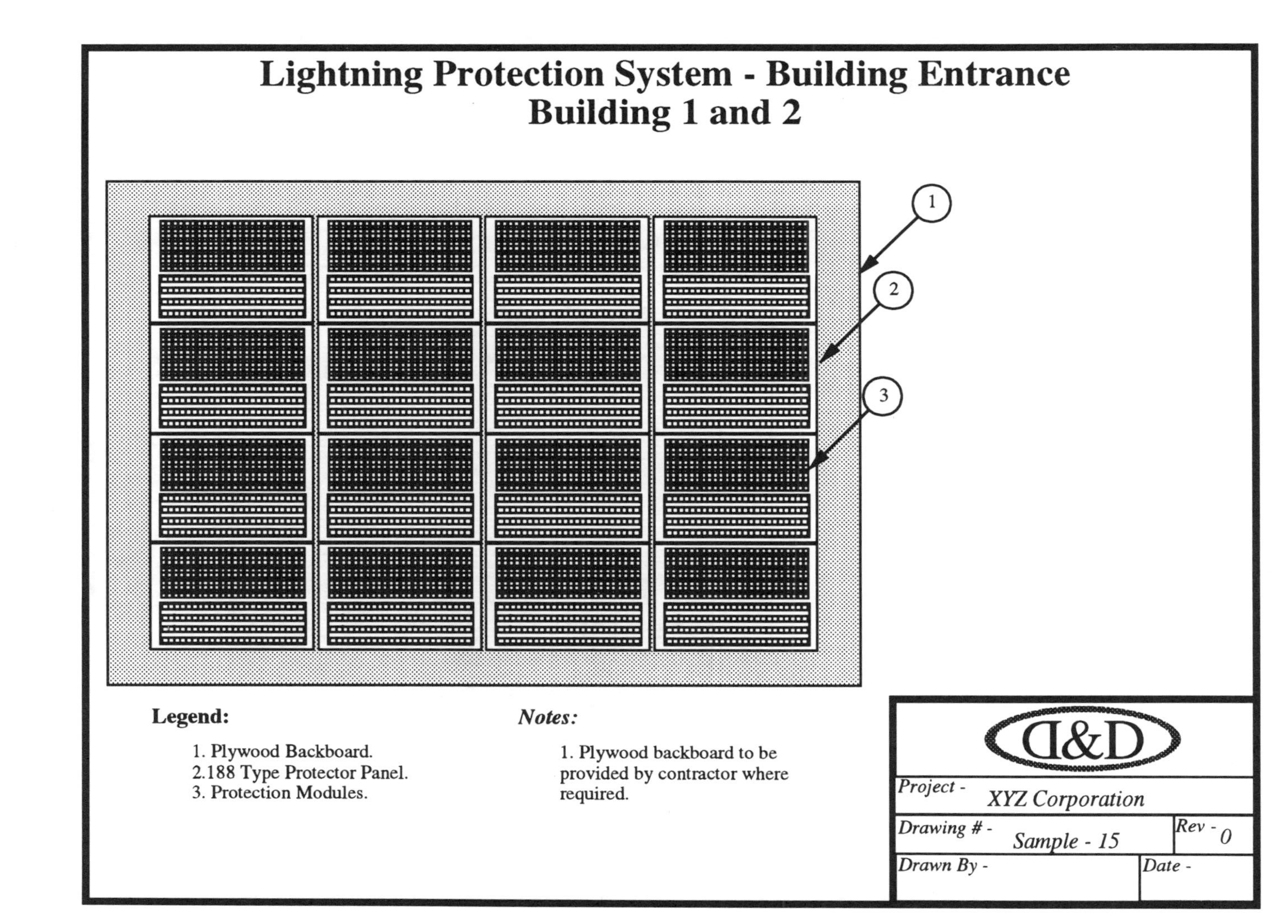

Lightning Protection System - Building Entrance
Building 1 and 2
1
2
3
Legend:
1. Plywood Backboard.
2.188 Type Protector Panel.
3. Protection Modules.
Notes:
1. Plywood backboard to be provided by contractor where required.
D&D
Project - XYZ Corporation
Drawing # - Sample - 15
Rev - 0
Drawn By -
Date -

Lightning Protection System - Building Entrance
Building 2 and 3

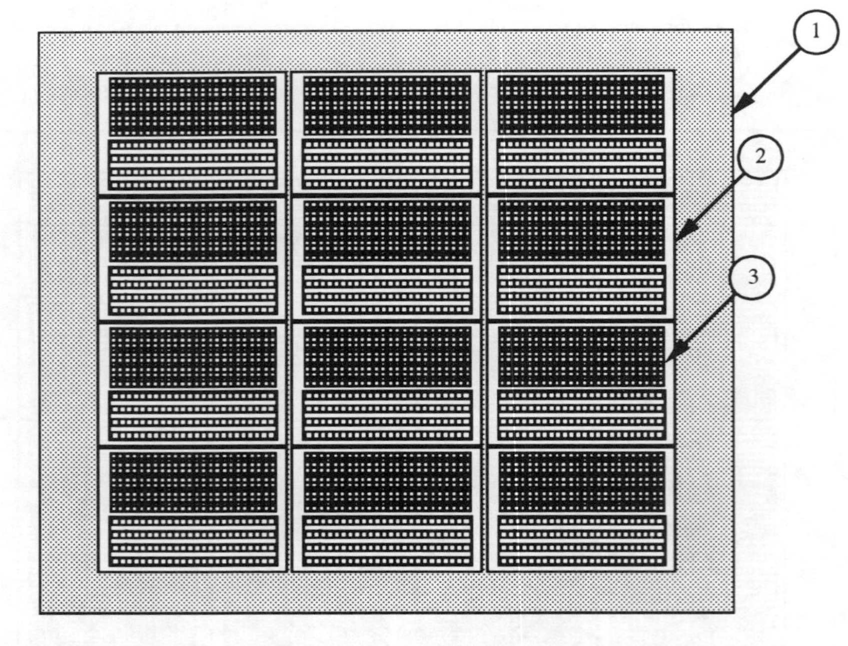

Legend:

1. Plywood Backboard.
2.188 Type Protector Panel.
3. Protection Modules.

Notes:

1. Plywood backboard to be provided by contractor where required.

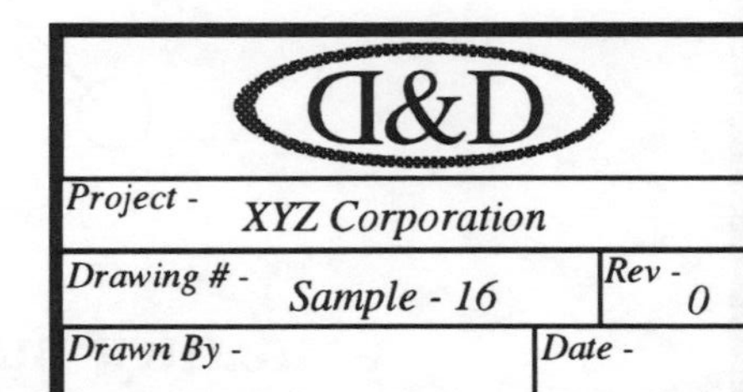

Project - XYZ Corporation	
Drawing # - Sample - 16	*Rev -* 0
Drawn By -	*Date -*

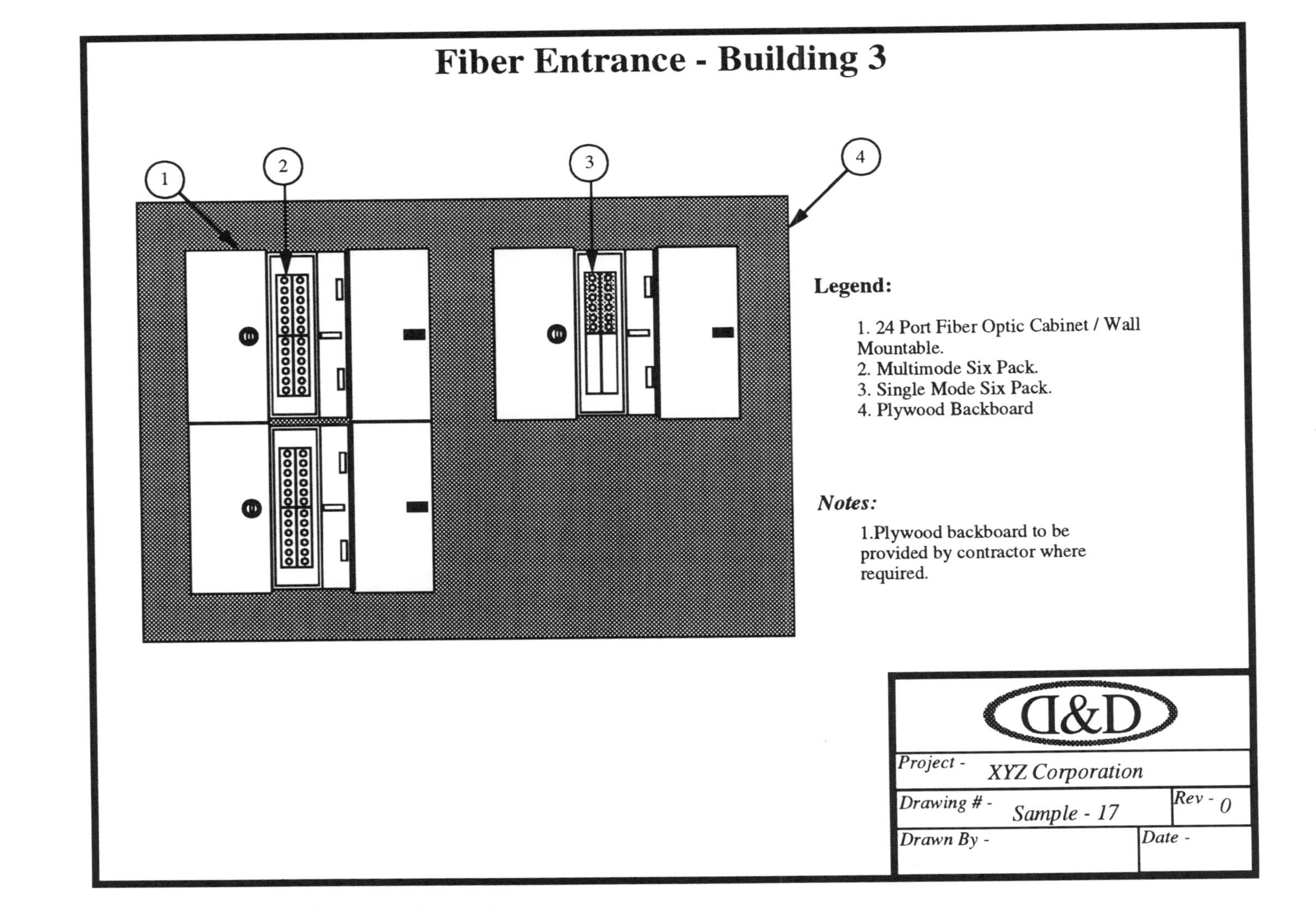
Fiber Entrance - Building 3
1
2
3
4
Legend:
1. 24 Port Fiber Optic Cabinet / Wall Mountable.
2. Multimode Six Pack.
3. Single Mode Six Pack.
4. Plywood Backboard
Notes:
1.Plywood backboard to be provided by contractor where required.
D&D
Project - XYZ Corporation
Drawing # - Sample - 17
Rev - 0
Drawn By -
Date -

Appendix 'B'

Pricing Format

Pricing Sheet - Building 1

Ref.	Description	Mfr.	Part #	Qty	Unit Mat.	Unit Labor	Unit Cost	Extended	Section Total
Section 1 Workstation Cabling									
1.1	Category 5 Plenum Station Cable			165000					
1.2	Category 3 Plenum Station Cable			82500					
1.3	4 Pair Data Terminations			1820					
1.4	4 Pair Voice Terminations			910					
1.5	Faceplates			455					
1.6	Surface Mount Boxes			250					
1.7	Dual Category 5 RJ45 Modules			455					
1.8	Dual Category 4 RJ45 Modules			455					
1.9	Blank Modules			455					
1.10	4 Pair Category 5 Certification Test			910					
1.11	Continuity & Polarity Test Voice Ports			910					
1.12	Label 4 Pair Cables			1365					
1.13	Label Workstation, Patch Panels & 110 Blocks			455					
Section 2 Distribution System									
2.1	100 Pair Category 3 Plenum Feeder Cable			4500					
2.2	Plenum Innerduct			700					
2.3	24 Strand Fiber Optic Cable			700					
2.4	8 Strand Fiber Optic Cable			50					
2.5	25 Pair Category 5 Tie Cable			1000					
2.6	Voice Feeder Pair Terminations			3200					
2.7	Multimode ST Connectors			112					
2.8	Singlemode ST Connectors			24					
2.9	25 Pair Category 5 Cable Terminations			150					
2.10	Test Voice Feeders (25 Pair Groups)			64					
2.11	Test Multimode FO (850nm & 1300nm)			56					
2.12	Test Single Mode FO (1300nm & 1550nm)			12					
2.13	Category 5 Certification Test Tie Cables			18					

Ref.	Description	Mfr.	Part #	Qty	Unit Mat.	Unit Labor	Unit Cost	Extended	Section Total
Section 3 - Termination Equipment									
3.1	19" Free Standing Racks			8					
3.2	48 Port Category 5 Patch Panels			22					
3.3	24 Port Category 5 Patch Panels			3					
3.4	Data Rack Wire Managers			30					
3.5	Voice Rack Wire Managers			6					
3.6	Vertical Wire Managers for Voice Racks			24					
3.7	32 Position Fiber Optic Cabinet			3					
3.8	72 Position Fiber Optic Cabinet			1					
3.9	Multimode six packs			20					
3.10	Single Mode Six Packs			5					
3.11	110C-4 Connecting Block			455					
3.12	110C-5 Connecting Block			640					
3.13	300 Pair 110 Blocks with Legs			4					
3.14	300 Pair 110 Blocks w/o Legs			12					
3.15	Cable Trough (feet)			20					
3.16	Plywood Backboard (4' x 8')			0					
Section 4 - Miscellaneous									
4.1	Ladder Rack (feet)			60					
4.2	Fire Stop Putty (bricks)			3					
4.3	Spiral Wrap (feet)			200					
4.4	Hardware (Cable Ties, Beam Clamps, etc.)			1 lot					
4.5	Mounting Bars (300pr blocks to 19" racks)			12					
Total Price Building 1									

Pricing Sheet - Building 2

Ref.	Description	Mfr.	Part #	Qty	Unit Mat.	Unit Labor	Unit Cost	Extended	Section Total
Section 1 Workstation Cabling									
1.1	Category 5 Plenum Station Cable			100000					
1.2	Category 3 Plenum Station Cable			50000					
1.3	4 Pair Data Terminations			1076					
1.4	4 Pair Voice Terminations			538					
1.5	Faceplates			269					
1.6	Surface Mount Boxes			150					
1.7	Dual Category 5 RJ45 Modules			269					
1.8	Dual Category 4 RJ45 Modules			269					
1.9	Blank Modules			269					
1.10	4 Pair Category 5 Certification Test			538					
1.11	Continuity & Polarity Test Voice Cables			538					
1.12	Label 4 Pair Cables			807					
1.13	Label Workstation, Patch Panels & 110 Blocks			269					
Section 2 Distribution System									
2.1	100 Pair Category 3 Plenum Feeder Cable			7000					
2.2	Plenum Innerduct			1000					
2.3	24 Strand Fiber Optic Cable			400					
2.4	8 Strand Fiber Optic Cable			0					
2.5	25 Pair Category 5 Tie Cable			400					
2.6	Voice Feeder Pair Terminations			7600					
2.7	Multimode ST Connectors			96					
2.8	Singlemode ST Connectors			24					
2.9	25 Pair Category 5 Cable Terminations			150					
2.10	Test Voice Feeders (25 Pair Groups)			304					
2.11	Test Multimode FO (850nm & 1300nm)			48					
2.12	Test Single Mode FO (1300nm & 1550nm)			12					
2.13	Category 5 Certification Test Tie Cables			18					

Ref.	Description	Mfr.	Part #	Qty	Unit Mat.	Unit Labor	Unit Cost	Extended	Section Total
Section 3 - Termination Equipment									
3.1	19" Free Standing Racks			8					
3.2	48 Port Category 5 Patch Panels			12					
3.3	24 Port Category 5 Patch Panels			3					
3.4	Data Rack Wire Managers			23					
3.5	Voice Rack Wire Managers			7					
3.6	Vertical Wire Managers for Voice Racks			24					
3.7	32 Position Fiber Optic Cabinet			4					
3.8	72 Position Fiber Optic Cabinet			2					
3.9	24 position Wall Mount FO Cabinet			0					
3.9	Multimode six packs			16					
3.10	Single Mode Six Packs			4					
3.11	110C-4 Connecting Block			269					
3.12	110C-5 Connecting Block			1520					
3.13	300 Pair 110 Blocks with Legs			8					
3.14	300 Pair 110 Blocks w/o Legs			14					
3.15	Cable Trough (feet)			20					
3.16	Plywood Backboard (4' x 8')			5					
Section 4 - Miscellaneous									
4.1	Ladder Rack (feet)			60					
4.2	Fire Stop Putty (bricks)			5					
4.3	Spiral Wrap (feet)			200					
4.4	Hardware (Cable Ties, Beam Clamps, etc.)			1 lot					
4.5	Mounting Bars (300pr blocks to 19" racks)			14					
Total Price Building 2									

Pricing Sheet - Building 3

Ref.	Description	Mfr.	Part #	Qty	Unit Mat.	Unit Labor	Unit Cost	Extended	Section Total
Section 1 Workstation Cabling									
1.1	Category 5 Plenum Station Cable			100000					
1.2	Category 3 Plenum Station Cable			50000					
1.3	4 Pair Data Terminations			1052					
1.4	4 Pair Voice Terminations			526					
1.5	Faceplates			263					
1.6	Surface Mount Boxes			150					
1.7	Dual Category 5 RJ45 Modules			263					
1.8	Dual Category 4 RJ45 Modules			263					
1.9	Blank Modules			263					
1.10	4 Pair Category 5 Certification Test			526					
1.11	Continuity & Polarity Test Voice Cables			526					
1.12	Label 4 Pair Cables			789					
1.13	Label Workstation, Patch Panels & 110 Blocks			263					
Section 2 Distribution System									
2.1	100 Pair Category 3 Plenum Feeder Cable			3000					
2.2	Plenum Innerduct			500					
2.3	24 Strand Fiber Optic Cable			500					
2.4	8 Strand Fiber Optic Cable			0					
2.5	25 Pair Category 5 Tie Cable			100					
2.6	Voice Feeder Pair Terminations			2400					
2.7	Multimode ST Connectors			96					
2.8	Singlemode ST Connectors			24					
2.9	25 Pair Category 5 Cable Terminations			50					
2.10	Test Voice Feeders (25 Pair Groups)			48					
2.11	Test Multimode FO (850nm & 1300nm)			48					
2.12	Test Single Mode FO (1300nm & 1550nm)			12					
2.13	Category 5 Certification Test Tie Cables			12					

Ref.	Description	Mfr.	Part #	Qty	Unit Mat.	Unit Labor	Unit Cost	Extended	Section Total
Section 3 - Termination Equipment									
3.1	19" Free Standing Racks			4					
3.2	48 Port Category 5 Patch Panels			14					
3.3	24 Port Category 5 Patch Panels			2					
3.4	Data Rack Wire Managers			18					
3.5	Voice Rack Wire Managers			0					
3.6	Vertical Wire Managers for Voice Racks			0					
3.7	32 Position Fiber Optic Cabinet			2					
3.8	72 Position Fiber Optic Cabinet			1					
3.9	24 position Wall Mount FO Cabinet			3					
3.9	Multimode six packs			16					
3.10	Single Mode Six Packs			4					
3.11	110C-4 Connecting Block			263					
3.12	110C-5 Connecting Block			480					
3.13	300 Pair 110 Blocks with Legs			9					
3.14	300 Pair 110 Blocks w/o Legs			0					
3.15	Cable Trough (feet)			30					
3.16	Plywood Backboard (4' x 8')			3					
Section 4 - Miscellaneous									
4.1	Ladder Rack (feet)			60					
4.2	Fire Stop Putty (bricks)			3					
4.3	Spiral Wrap (feet)			200					
4.4	Hardware (Cable Ties, Beam Clamps, etc.)			1 lot					
4.5	Mounting Bars (300pr blocks to 19" racks)			0					
Total Price Building 3									

Pricing Sheet - Outside Plant Cabling

Ref.	Description	Mfr.	Part #	Qty	Unit Mat.	Unit Labor	Unit Cost	Extended	Section Total
Section 1 Cabling									
1.1	600 Pair Foam Filled Qualpeth Cable			1400					
1.2	400 Pair Foam Filled Qualpeth Cable			350					
1.3	48 Strand Outside Plant FO Cable			700					
1.4	12 Strand Outside Single Mode			700					
1.5	Innerduct			700					
1.6	Copper Terminations (to Lightning Protection)			5600					
1.7	Multimode ST Connectors			192					
1.8	Singlemode ST Connectors			48					
1.9	Test Voice Feeder (25 Pair Groups)			112					
1.10	Test Multimode FO (850nm & 1300nm)			96					
1.11	Test Single Mode FO (1300nm & 1550nm)			24					
1.12	Through put test FO @1300nm			120					
1.13	Label Cables			9					
1.14	Label Punch Blocks & Patch Panels			9					
Section 2 - Termination Equipment									
3.1	Lightning Protection Panels			56					
3.2	Protection Modules			5600					
3.3	72 Position Fiber Optic Cabinet			0					
Section 4 - Miscellaneous									
4.1	Ladder Rack			40					
4.2	Fire Stop Putty (bricks)			5					
4.3	Hardware (Kellems,Pulling Soap etc.)			1					
4.4	Grounding of Lightning Protection			1					
4.5	Plywood Backboard			4					
Total Price Outside Plant Cabling									

Cable Plant Summary

SECTION	TOTAL COST
Building 1	
Building 2	
Building 3	
Outside Plant Cabling	
Total Base Project	

Appendix 'C'

Unit Pricing

Unit Prices

Ref.	Description	Mfr.	Part #	Qty	Unit Mat.	Unit Labor	Unit Cost	Extended	Section Total
Section 1 Unit Workstation Cabling									
1.1	Category 5 Plenum Station Cable			360					
1.2	Category 3 Plenum Station Cable			180					
1.3	4 Pair Data Terminations			4					
1.4	4 Pair Voice Terminations			2					
1.5	Faceplates			1					
1.6	Surface Mount Boxes			1					
1.7	Dual Category 5 RJ45 Modules			1					
1.8	Dual Category 4 RJ45 Modules			1					
1.9	Blank Modules			1					
1.10	4 Pair Category 5 Certification Test			2					
1.11	Continuity & Polarity Test Voice Cables			2					
1.12	Label 4 Pair Cables			6					
1.13	Label Workstation, Patch Panels & 110 Blocks			1					
Section 3 - Cross Connects									
3.1	1 Pair Cross Connect			1					
3.2	2 Pair Cross Connect			1					
Section 3 - Patch Cables									
4.1	2' Category 5 PVC Patch Cords			1					
4.2	3' Category 5 PVC Patch Cords			1					
4.3	4' Category 5 PVC Patch Cords			1					
4.4	8' Category 5 PVC Patch Cords			1					
4.5	14' Category 5 PVC Patch Cords			1					
4.6	4' Duplex ST to ST Jumper			1					
4.7	5' Duplex ST to ST Jumper			1					
4.8	6' Duplex ST to ST Jumper			1					

Appendix 'D'

Labor Rates

Provide hourly labor rates for straight time, over time and double time conditions and define the periods of the day when these rates are effective.

Straight Time Rate	**$0.00 per m/h**

Effective Hours -

Over Time Rate	**$0.00 per m/h**

Effective Hours -

Double Time Rate	**$0.00 per m/h**

Effective Hours -

Appendix 'E'

Invoicing Format

Invoice Format - Building 1

Ref.	Description	Qty	Previous Complete	Current Complete	Unit Mat.	Unit Labor	Unit Cost	Extended	Previous Billed	Current Billed
Section 1 Workstation Cabling										
1.1	Category 5 Plenum Station Cable									
1.2	Category 3 Plenum Station Cable									
1.3	4 Pair Data Terminations									
1.4	4 Pair Voice Terminations									
1.5	Faceplates									
1.6	Surface Mount Boxes									
1.7	Dual Category 5 RJ45 Modules									
1.8	Dual Category 4 RJ45 Modules									
1.9	Blank Modules									
1.10	4 Pair Category 5 Certification Test									
1.11	Continuity & Polarity Test Voice Cables									
1.12	Label 4 Pair Cables									
1.13	Label Workstation, Patch Panels & 110 Blocks									
Section 2 Distribution System										
2.1	100 Pair Category 3 Plenum Feeder Cable									
2.2	Plenum Innerduct									
2.3	24 Strand Fiber Optic Cable									
2.4	8 Strand Fiber Optic Cable									
2.5	25 Pair Category 5 Tie Cable									
2.6	Voice Feeder Terminations									
2.7	ST Connectors									
2.8	25 Pair Category 5 Cable Terminations									
2.9	Continuity & Polarity Test Voice Feeders									
2.10	Power Meter Test FO Cable (850nm)									
2.11	Category 5 Certification Test Tie Cables									

Ref.	Description	Qty	Previous Complete	Current Complete	Unit Mat.	Unit Labor	Unit Cost	Extended	Previous Billed	Current Billed
Section 3 - Termination Equipment										
3.1	19" Free Standing Racks									
3.2	48 Port Category 5 Patch Panels									
3.3	24 Port Category 5 Patch Panels									
3.4	Wire Managers									
3.5	24 Port Fiber Optic Patch Panels									
3.6	24" 110 Block Mounting Frames									
3.7	110C-4 Connecting Block									
3.8	110C-5 Connecting Block									
3.9	300 Pair 110 Blocks									
3.10	Cable Trough									
3.11	Plywood Backboard									
Section 4 - Miscellaneous										
4.1	Ladder Rack									
4.2	Fire Stop									
4.3	Spiral Wrap									
4.4	Hardware (Cable Ties, Beam Clamps, etc.)									
4.5	Mounting Bars (300pr blocks to 19" racks)									
Total Price										
Total Previously Billed										
Total This Billing										

Appendix 'F'

Workstation Schedule

Workstation Schedule

Floor/Bldg	Building 1	Building 2	Building 3	Totals
Basement	0	19	0	19
1st	91	119	102	312
2nd	212	125	167	504
3rd	152			152
Totals	455	263	269	987